SIGNS OF THE GODS?

ERICH VON DÄNIKEN

Translated by Michael Heron

SOUVENIR PRESS

First published in Germany by Econ Verlag
under the title *Prophet der Vergangenheit*

First Published in Great Britain 1980 by Souvenir Press Ltd,
43 Great Russell Street, London WC1B 3PA
and simultaneously in Canada

Reissued in paperback 1995
Reprinted 1997

ISBN 0 285 62435 0 (casebound)
ISBN 0 285 63270 1 (paperback)

Printed Photolitho in Great Britain by
Ebenezer Baylis and Son Limited
The Trinity Press, Worcester and London.

In memoriam
Rolf R. Bigler
who accompanied me
as friend
and
critic

Contents

Photographic Credits

With the exception of the Manna Machine (Dale-Sassoon, London, 1978), 'A global football network' (Sputnik, 9/1974) and King List WB 444 (Schmidtke: *Aufbau der babylonischen Chronologie*, Münster 1952), all photos and illustrations are by Erich von Däniken, who also holds the exclusive copyright.

1: In Search of the Ark of the Covenant

Agatha Christie, that incomparable detective-story writer, once gave an interview in which she explained the ideal pattern for a good crime story. She said that only when suspicion was clearly proved so that a criminal with a sound motive could be revealed at the end was a story satisfying and exciting. But she added that the working out of the plot was only convincing if it left lingering doubts *after* the end of the story. Agatha Christie was talking about fictitious crime. I want to tell you about a crime which actually took place but nevertheless fulfils all the conditions which the grand old lady laid down for a first-class detective story.

For me the crime began in a religion class. We were told that God commanded Moses to build an ark. We can read the instructions he was given in Exodus 25, 10. However they cannot have been purely verbal; he must have had a model of the ark:

> 'And look that thou make them after the pattern, which was shewed thee on the mount.' Exodus 25, 40

This ark is the material evidence involved in our crime. We must not lose sight of it. But although the crime is so incredibly old, experts are still arguing about it—just as nowadays the family quarrels about the presumptive author of a crime in a TV series.

What kind of object was the ark?

Theologians, who at this point in the story assume the functions of police investigators, have very contradictory ideas about it. In Pierers Universal-Lexikon (1) the ark, also known as the ark of the covenant, is described as:

'An acacia-wood chest, 1.75 m long, 1 m high and 1 m wide, covered inside and out with gold.'

Professor Dr Hugo Gressmann (2), the well-known theologian, makes the chest quite a bit smaller:

'About 1.25 m long, 0.75 m high and 0.75 m wide.'

Precise, but scanty details. We learn a good deal more in the Zohar, the most important of the Cabbalistic works, whose investigators showed more curiosity. In spite of the detailed information, the Zohar does not seem to have been included in the official reports, possibly because it was a secret Jewish work which appeared between A.D. 130–170. Yet it devotes nearly 50 (!) pages to the ark of the covenant and even gives minute details that escaped other criminologists.

At first glance it seems surprising that the 'ark of the covenant' is introduced as the 'Ancient of Days' in the Zohar, but a second glance makes it quite clear that the description refers to the ark.

The instructions in the Zohar are identical with the account in Exodus. Yahweh, the god of Israel, orders

The customary idea of how the ark of the covenant was transported. From an old bible.

Moses to build a chest for the 'Ancient of Days', giving precise details. This container, together with the remarkable 'Ancient of Days', is to be taken on the journey through the desert.

So far no one has denied that the ark existed, but the experts differ about the measurements. The theological criminal bureau also argues about the purposes of the mysterious chest.

Reiner Schmitt (3) claims that the ark was: 'a container for a sacred stone.'

Martin Dibelius (4) contradicts him and says it was: 'A mobile empty divine throne' or 'a travelling divine car on which a divinity stood or sat'.

How are we going to find out more about our material evidence if the theologians are not even unanimous about what it was used for?

The conviction that the ark of the covenant was empty because God lived in it, which the theologian R. Vatke threw into the debate in 1835, would make further research superfluous.

Harry Torczyner (6) stated that the ark of the covenant contained protocols, including two of the tablets of the law given to Moses. Here Harry Torczyner parts company with his colleague Martin Dibelius (4), who not only queries the accepted concept of ark of 'the covenant', but also doubts if Moses' tablets of the law were ever in the receptacle.

Investigation of the weight of the mysterious ark is equally confusing. The prophet Samuel, who was also a judge and must have been an accurate observer by profession, writes:

> 'Now therefore make a new cart, and take two milch kine, on which there hath come no yoke . . . And take the ark of the LORD, and lay it upon the cart: and put the jewels of gold, which ye return him for a tresspass offering, in a coffer by the side thereof.' I Samuel 6, 7–8

Judge Samuel even tells us about another cart for transporting the ark:

'And they set the ark of God upon a new cart, and brought it out of the house of Abinadab, that sitteth upon a hill.
And Uzza and Ahio, the sons of Abinadab, drove the new cart.' II Samuel 6, 3

'And . . . when they that bare the ark of the LORD had gone six paces

In spite of transport on one or two carts and the pulling power of two grown cows, the weight can hardly have exceeded 300 kg, for it was sometimes carried by the Levites, the priests in Yahweh's sanctuaries:

'And it was so, that when they that bare the ark of the LORD had gone six paces, he sacrificed an ox and a fatling' II Samuel 6, 13

Really we ought to lecture the biblical criminologists about their disagreements, as Moses did in Leviticus, when he gave the priestly Levites their instructions.

Lazarus Bendavid (1762–1832), a philosopher and mathematician, lived in Berlin. He was head of the Jewish Free School and a very open-minded man. He was also editor of the celebrated *Spenerschen Zeitung* (7). Contemporaries of Bendavid say that he was a well-known Jewish scholar and philosopher, who had come to the conclusion,

> 'that the ark in Moses' day contained a fairly complete system of electrical instruments and produced effects by them.'

Lazarus Bendavid was not only an intelligent man, he was also far ahead of his time. As an orthodox Jew had he read the Zohar? Did he come across the 'Ancient of Days' in it? Did it make him suspicious? Was he dissatisfied with the research then known to him? Of course he knew that only a fixed, clearly defined group of people was allowed access to the ark and that even the high priests could not visit it every day. Because the ark was dangerous!

Bendavid:

> 'According to the Talmudists mortal danger must always have been associated with these visits to the holy of holies.
>
> The high priest always entered with a certain fear and thought it was a good day if he came back safely.'

The crime story gets more complicated! The ark of the covenant changed hands! After a victorious war, the Philistines, a Hebrew tribe of western origin, confiscated the ark of the Lord. They had noticed how important the mysterious apparatus was to the Israelites and hoped to derive advantages from owning it.

But the Philistines had no instructions for use; they were quite unfamiliar with it. At any rate, in a very short time they realised that anyone who came into close contact with the ark fell sick or died. They began to pass the apparatus they had requisitioned from town to town, like a hot potato, but everywhere it was the same story. All who were curious enough to approach the dubious booty were stricken with boils and scales, and their hair fell out. Children and adults, all were overcome by excruciating vomiting and many perished in a horrible way.

Judge Samuel was a witness:

'They sent therefore and gathered together all the lords of the Philistines and they said, Send away the ark of the God of Israel, and let it go again to its own place, that it slay us not or our people: for there was a deadly destruction throughout all the city; the hand of God was very heavy there. And the men that died not were smitten with the emerods: and the cry of the city went up to heaven.' I Samuel 5, 11–12

The Philistines were in possession of the devilish object for seven months. Then all they wanted was to get rid of their booty. They put the chest on a cart, harnessed two cows to it and whipped the bellowing beasts to the borders of Beth-shemesh.

In the morning, when the men of Beth-shemesh came into the valley to harvest the wheat, they saw the cart with the ark on it. They slaughtered the cows on the spot and then summoned the Levites, who alone knew how to handle the ark. But 50,070 people, who had no idea how dangerous the ark was, died in a horrible way. They were curious and came too near the ark, and the Lord smote them.

I Samuel 6, 19

Now the ark was back in the hands of the people who built it, but we still do not know what this apparatus really was.

The crime story continues, but a solution suggests itself.

In 1978, a book called *The Manna Machine*, the joint work of George Sassoon, an electronics consultant, and Rodney Dale, a biologist and engineering writer, was published in London. These British investigators used the Zohar's very precise description of the 'Ancient of Days', which they interpreted and reconstructed in the light of present-day technical and biological knowledge. They decided that the ark of the covenant was actually a technical machine—as Bendavid also assumed—that supplied the Israelites on their trek through the desert with a food rich in protein, namely manna.

Now the enquiry has taken a giant step forward. The ark of the covenant = the Ancient of Days = the manna machine. A formula as faultless as the one times table!

As technology is not exactly a special concern of theologians, we can remove them from the team of criminologists. It is now clear:
that the ark of the covenant was *not* the holy of holies, but the receptacle for a machine which produced food;
—that it could only be approached by the 'elect', i.e. those who were trained to operate it;
—that the curious who were rash enough to approach it fell ill or died, because the machine was highly radioactive.

On the basis of the information we now have, the 'Case of the Ark of the Covenant' looks as follows:

For unknown reasons extraterrestrials were interested in separating a group of people from their environment and cutting them off from all contact with the 'rest of mankind' for two generations. Through their intermediary, a prophet, they ordered the withdrawal from civilisation of the chosen group. Moses—it may also have been another of the elect—led the Israelites through the wilderness. The extraterrestrials saved the nomadic people from their enemies, for the attacking Egyptians were drowned:

'And the waters returned, and covered the chariots, and the horsemen, and all the host of Pharaoh that came into the sea after them; there remained not so much as one of them. Exodus 14, 28

The FBI, like any other group of qualified investigators, would dismiss the theories put forward by theologians to explain the above phenomenon as absolute nonsense. One of them claims that the Israelites marched through a sea thick with reeds or sand banks at ebb-tide, while the pursuing Egyptians were surprised by the waters of the flood-tide.

We can credit the chosen people with many special gifts, but we cannot imagine that the Egyptians, who first divided the year into 365 days after observing the regular advent of the Nile floods, did not know just as much about ebb- and flood-tides as the Israelites.

No, the Egyptians did not rush blindly to their destruction! They were *deliberately* led astray by a mysterious 'angel'—by a pillar of fire:

'And *the angel of God*, which went before the camp of Israel, removed and went *behind them*; and the pillar of the cloud went from before their face, and stood behind them. And it came between the camp of the Egyptians and the camp of Israel; and it was a cloud and darkness to them, but it gave light by night to these: so that the one came not near the other all the night.' Exodus 14, 19-20

This cloud was not a chance meteorological phenomenon, as some scholars have suggested. Moses expressly says that the 'pillar of cloud' was a signal to lead the Israelites:

'And the LORD went before them by day in a pillar of cloud, to lead them the way; and by night in a pillar of fire, to give them light; that they might go by day and night. He took not away the pillar of the cloud by day, nor the pillar of fire by night, from before the people.'

Exodus 13, 21–22

Chance meteorological phenomena usually appear for seconds, minutes or even hours, but not for months and years. This explanation simply does not stand up to examination.

As we are not on the trail of an individual Israelite or a small group, we have a slightly easier task than criminologists who have to try to uncover a narrow trail. Ahead of us lies the broad path left by a giant migration that wound slowly through the wilderness. The enemy were destroyed; the road was clear. Nevertheless, it was a tremendous undertaking to lead thousands of men, women and children through territory where it was impossible to live on wild fruit or game. Even modern armies have come to grief for lack of fresh provisions.

Temperatures in deserts, with their hostile environment, vary between 58°C and –10°C. Annual rainfall barely averages 10 cm. There are no natural products that could satisfy the hunger of a giant host, yet Moses risked marching his people through the endless burning wilderness.

Who supplied the people of Israel with food?

Extraterrestrials helped them and Moses knew it. For 'the

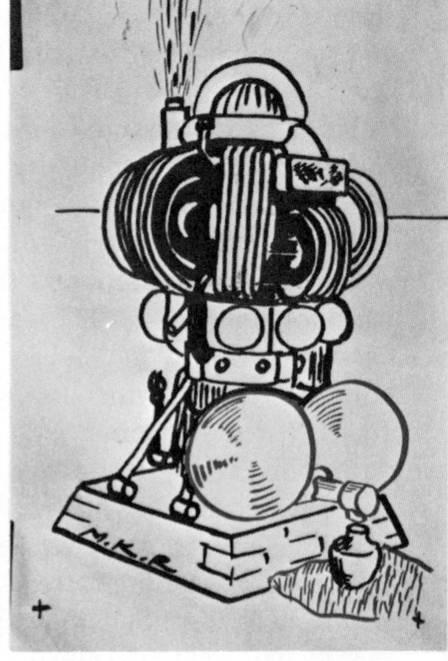

The British scientists George Sassoon and Rodney Dale reconstructed the 'ancient of days' from a description in the book Zohar. The result was a machine which produced a food rich in protein, the manna of the desert, from green algae subjected to radiation.

Lord' who appeared to him in the burning bush showed him a machine that would take care of any worries about food supplies during the long years of the exodus.

It was a marvellous machine. It stored up water from the night dew, then mixed it with a microscopic type of green algae (chlorella) and produced as much food as was needed. There were complaints about having the same menu every day, but no one had to go hungry, as Moses pointed out when the people rebelled.

The food synthesised from dew and green algae was produced by radiation. Radiation needs energy. Where could it be obtained in the barren desert? What kind of source of energy was it that was unexhausted after 40 years?

Today we can erase the question mark. From our knowledge of existing technology it can only have been a mini-nuclear reactor. Reactors of this type exist and have been in service for some time. We have known this since February, 1978—from official sources.

The Russian espionage satellite 'Cosmos 954' crashed in the countryside around the Great Slave Lake in Canada. The Strategic Bomber Command of the US Air Force was alerted. Coded messages and orders went out to submarines at sea. Leave was stopped for all crews at rocket stations. The hot lines at NATO were glowing. 'Cosmos 954' was carrying more than 45 kg of radioactive uranium 235, a source of energy which, according to experts, could radiate for 1000 years or more and contaminate country and people with poisonous radioactive clouds. When 'Cosmos 954' crashed, the reactor melted owing to frictional heat and its deadly contents were released. After the politicians had shaken hands politely, reassurances were given. But their handshakes did not remove the radioactive emissions, they merely defused the explosive political situation.

Soon afterwards the Indian government revealed that years before the CIA had arranged for expert mountain climbers to take a mini-reactor to the Himalayas where it would act as an inexhaustible source of energy for the devices which are listening in to China around the clock.

Mini-reactors release energy by the disintegration of plutonium. Radiation energy is transformed directly into electricity, a process which differs from that in large atomic power stations which employ heavy water and fuel rods. The mini-reactor gives off radiation. It is dangerous, but not fatal provided you do not stay close to it for too long. After all, the valiant mountaineers carried one to the Himalayas and came back safely.

Mini-reactors will be an indispensable source of energy in the spaceships of the future. A machine that uses radiation to produce a protein-rich foodstuff from water and green algae will also be of the greatest importance for interstellar space travel. I am certain that the discoveries of Sassoon and Dale have already been closely examined by space travel experts. With a manna machine on board the problem of a basic foodstuff for space travellers is solved.

Obviously the ark which 'the Lord' showed Moses on the sacred mount could not be left out in the open. Perhaps it had to be protected from desert sandstorms, perhaps the very high daytime temperatures were harmful to it, perhaps too, the wandering Israelites were not supposed to know about the mysterious factory which produced their food. At any rate, an ark, a beautifully made chest, was built around it, following the model and instructions given. Therefore the ark of the covenant was not the manna machine, but merely the container in which it was preserved and transported. The two-fold goal was reached. The sensitive machine was protected from both damaging outside influences and inquisitive glances. During long rest periods a tent was erected around 'the factory'. It was never set up inside the camp, because of the dangerous radioactivity:

'Now Moses used to take the tent and to pitch it *outside the camp*, far off from the camp, and he called it the tent of meeting.' Exodus 33, 7

Let us keep on the trail of our material evidence. We already know quite a lot about it, including how it worked.

When Sassoon and Dale reconstructed the machine, following the Book of Zohar, they learnt that the 'Ancient of Days' indefatigably turned out the glutinous foodstuff, manna, for six consecutive days. A six-day week. The machine was cleaned on the seventh day. This maintenance work was performed by the Levites, who had been instructed by Aaron, Moses' brother. Aaron had climbed the mount with Moses and had obviously been given a crash course. The 'Lord' had instructed him:

'And the LORD said unto him, Away, get thee down, and thou shalt come up, thou and Aaron with thee: but let not the priests and the people break through to come up unto the LORD, lest he break forth upon them.'

Exodus 19, 24

What conclusions can be drawn at this stage of the the investigation?

—the extraterrestrials escorting the wandering people wanted to separate a group from their environment;

—the extraterrestrials did not possess a fleet of shuttle craft, otherwise they would have transported their protegés by spaceship;

—the extraterrestrials who landed were few in number. When their spaceship touched down on the mount, the Commander expressly ordered Moses to have a fence built round the mountain so that no one could break in:

'And the LORD said unto Moses, Go down, charge the people, lest they break through unto the LORD to gaze and many of them perish . . . And Moses said unto the LORD, the people cannot come up to mount Sinai: for thou didst charge us, saying, Set bounds about the mount, and sanctify it.' Exodus 19, 21-23

The small group of extraterrestrials demonstrated their superiority with technological tricks—the pillar of fire that

Moses faces the ark of the covenant in the 'tabernacle of the tent of the congregation'.

guided the Israelites and the drowning of the Egyptian army; the spaceship's drive unit emitted burning hot gases and made a terrifying noise:

'And mount Sinai was altogether on a smoke, because the LORD descended upon it in fire: and the smoke thereof ascended as the smoke of a furnace, and the whole mount quaked greatly.' Exodus 19, 18

A food-producing machine was unloaded from the spaceship and handed over to Moses and Aaron;

—when the machine was transported it was put in a container, the ark;

—the machine was carried on a cart yoked to oxen, but cannot have weighed more than 300 kg, because it was occasionally carried by men with poles on their shoulders;

—people who rashly came too near the ark fell ill, died or were smitten with boils, scales and pustules;

—no one knew what was transported in the ark. The people only knew that 'the Lord' supplied them with food. The tabernacle in which the ark stood concealed a secret;

—the specially trained Levites, wearing protective clothing, looked after its maintenance, but even they did not know what kind of machine it was. They were afraid of it and with good reason, for there were accidents in which priests, too, were killed.

That is as much as we can say at the present stage of our investigation of the 'Case of the Ark of the Covenant'.

What happened next?

What became of the ark and its mysterious contents?

Where did it end up?

Is it still in existence?

Can we find it again? How?

A monstrosity of that size and weight cannot have vanished into thin air.

Let us continue our tracking.

We can conclude from the description in Exodus that the machine functioned as long as it was serviced regularly. After the return to the promised land it was no longer necessary, for milk and honey flowed there—a change in the monotonous menu at last.

However, there were probably rumours that the home-comers had taken on their trek a strange object that supplied them with food. Industrial espionage began. Every monarch wanted to possess the tireless machine. We have noted the Philistines' victorious battle against the Israelites and the capture of the machine, which they returned after a series of disasters.

What was the fate of the ark when it was deposited in Beth-shemesh?

It was kept in a hut for at least twenty years:

'And the men of Kirjath-jearim came, and fetched up the ark of the LORD, and brought it into the house of Abinadab in the hill, and sanctified his son to keep the ark of the LORD.

And it came to pass, while the ark abode in Kirjath-jearim, that the time was long; for it was twenty years:

and all the house of Israel lamented after the LORD.'

I Samuel 7, 1–2

Obviously the machine was no longer working; no one bothered about it and it was completely forgotten.

It was Saul, the first king of Israel, who lived about 1000 B.C., who reminded his son-in-law, King David (1013–973 B.C.), about the ark that had aroused so much attention in its day. When David began to take an interest in the mysterious object, it was still in Arbinadab's hut, just as it had been delivered to him. His curiosity was aroused, but it did not occur to him to transfer the ark to a worthy home in the palace he was just having built. Perhaps he was frightened by the hair-raising stories which circulated about it. Perhaps he did not consider the monster important enough to build a special room for it. In any case, it was quite a while before he followed his royal father-in-law's tip and went with 30,000 men to 'Baale of Judah, to bring up from thence the ark of God.' II Samuel 6, 1

During the removal another spectacular accident happened: 'And they set the ark of God upon a new cart, and brought it out of the house of Abinadab that was in Gibea: and Uzza and Ahio, the sons of Abinadab, drove the new cart and Ahio went before the ark . . . And when they came to Nachon's threshing floor, Uzza put forth his hand to the ark of God, and took hold of it; for the oxen shook it. And the anger of the LORD was kindled against Uzza; and God smote him there for his error; and there he died by the ark of God.' II Samuel 6, 3–7

A new hint in our search for the machine. It still gave electric shocks after twenty years of disuse! In other words, the mini-reactor was still radiating energy. Very important for our further pursuit of the ark.

Minor mishaps were overcome and the ark and its contents reached Jerusalem safely. King David was so happy about it that he danced for joy. He took off his clothes and leapt about stark naked. The joy of ownership? Or did he hope to please Yahweh so that he would set the machine going again? Did he want to conjure up manna for his people?

Although he was proud of having the ark, David would not agree to keep it in his palace, nor did he have a temple built for it:

'And they brought in the ark of the LORD, and set it in his place, in the midst of the tabernacle that David had pitched for it.' II Samuel 6, 17

Once again the mysterious object was shrouded in silence until David's successor, King Solomon (about 965–926 B.C.), had the ark placed in the holy of holies, a specially protected room in the temple. There it remained inviolate for more than 300 years of wars and disasters in the kingdom of Israel. During this period of time looters stole precious stones and gold from the temple on at least four occasions, but the ark was untouched. At any rate, it is no longer mentioned in the chronicles. Looters also took less precious artefacts than jewels. Had they no idea of the ark's existence? Were they afraid of its mysterious contents? Had the Israelites hidden it as a highly-prized souvenir of their desert trek? Did no one know where it was? Is that the reason why its traces were obliterated for so long? At all events we can deduce from another passage in the Bible that little importance was attached to it:

'Put the holy ark in the house which Solomon the son of David king of Israel did build: it shall not be a burden upon your shoulders.' II Chronicles 35, 3

It is suspected that the ark got lost during the destruction of Jerusalem (586 B.C.). We must also follow up this hint, however complicated the investigation may be. We must not throw in the sponge too soon.

But first a second progress report:

—The machine no longer produced manna.

—No one knew how to work it.
—In spite of a long time in store, the mini-reactor still functioned. The electric current it produced was strong enough to kill Uzza on the spot when he touched the ark.
—Three kings, Saul, David and Solomon were afraid of the ark and concealed it.
—With the passage of time the ark lost the religious significance it had had during the journey through the wilderness.
—The extraterrestrials had obviously disappeared.

Let us follow the new trail.

During the lifetimes of the prophet Jeremiah (627–585 B.C.) and his contemporary Ezekiel, the extraterrestrials suddenly reappeared. They asked Jeremiah to get rid of the machine, which was still dangerously radioactive.

Jeremiah, one of the great prophets of the Old Testament, was a prickly character. He grew up in the small town of Anathoth, north of Jerusalem, and came from a priestly family. He became most unpopular with his contemporaries, because he railed against idolaters, summoning them to repentance and lashing out at every kind of immorality. In short he held up to the people a mirror in which they reluctantly recognised themselves. Like all prophets, he was also a politician with a flair. He prophesied Israel's downfall and the destruction of the temple in Jerusalem.

It is understandable that Jehoiakim, the king of Judah (608–598 B.C.), took little or no pleasure in Jeremiah's words. This did not stop him, at the beginning of Jehoiakim's reign, from standing in the forecourt of the temple and making an inflammatory speech in which he hurled his thunderbolts at the congregation. Jeremiah was an awkward customer and so people tried to lay traps for him and silence him completely.

In the midst of all this intrigue, the astute Jeremiah had a bright idea. In 605 B.C. he got his pupil Baruch to write down his words and spread them abroad. A year later, during a religious fast, Baruch read Jeremiah's words to the people assembled in the temple. The officials foamed with rage and spoke of incitement to rebellion against king Jehoiakim. They took Baruch's rolls from him and gave them to the king. He was equally angry, cut up the rolls with a knife and threw the pieces into the fire.

From then on, Jeremiah and Baruch had to go undercover.

The prophets did not confine themselves to religious themes.

They were out and out politicians and demagogues. When they spoke, they attacked topics of current political interest. And how! They were masters of rhetoric and knew when the people were ready to kick against the pricks.

King Jehoiakim and later King Zedekiah were looked on as Egyptian vassals. Jeremiah, however, was on the side of the Chaldeans (Babylonians) and therefore anti-Egyptian. Jehoiakim and Zedekiah allowed heathen customs which were on the increase in Israel. Jeremiah was outspoken about this immorality. He might easily have incited his fellow-countrymen to rebellion, for the Israelites had to pay tribute in those days. The Israelite king saw his chance, made an alliance with the Egyptians and suspended reparation payments.

Nebuchadnezzar II, king of the Chaldeans (605–562 B.C.), would not put up with this insult. He sent an army from Syria to besiege Jerusalem, which he captured in 597 B.C.

In these dire straits, Zedekiah sent a messenger to the hated Jeremiah. He had no comfort for the king, only the disheartening advice to submit unconditionally to the Babylonians.

An Egyptian army appeared as if by magic and took a hand in the battle. Suddenly the Babylonians had to defend themselves against both the Israelites and the Egyptians. Superficially it looked as if the wily Jeremiah had erred badly in his prognosis, but only superficially, for the Babylonians defeated the Egyptians decisively and went back to besiege Jerusalem.

At no time in history have rulers been happy when an outsider proves right. Methods of punishment differed, but punished they always were. Either they or their reputations were destroyed. Jeremiah's enemies at court persuaded Zedekiah to kill him. The king had the prophet and politician lowered into a deep dungeon, the bottom of which was full of deep noxious mud. The troublesome prophet was left there to starve to death.

In any good crime story the 'hero' is always saved at the last minute in some unexpected way. Jeremiah had the same good fortune!

One of King Zedekiah's advisers was a young Ethiopian called Ebed-melech. Thanks to his enormous influence over the monarch, he managed to get the starving and freezing Jeremiah pulled up out of the dungeon (9).

Jerusalem did not last long. The Babylonians breached the city walls; King Zedekiah was taken prisoner and his eyes were put out, 10,000 Israelites went into exile . . .

'. . . all the army leaders and men who could bear arms, and the locksmiths and smiths, only the lowly remained behind. The temple and royal palace treasures were also taken to Babylon, and Solomon's golden vessels were broken up in the temple itself.' (10)

At last Jeremiah regained his freedom!

But the question remains: where was the ark of the covenant? Criminal cases like this are not easy to solve. You have to follow many trails before you find the right one. So even if we may have to make detours, we must stay on the trail of our super machine.

Let us take a shift in time.

Jerusalem was captured by the Babylonian King Nebuchadnezzar in 597 B.C. His son Belshazzar ruled about the middle of the sixth century B.C. And then a mysterious thing happened.

King Belshazzar had invited 1000 guests to a great feast. In a bacchanalian mood, he ordered the gold and silver vessels which his father had taken from Jerusalem to be filled and carried into the hall, amid great exultation. Drunk with the sweet wine, the boisterous guests grabbed at the sacred objects. Great fun, that idea of Belshazzar's!

In the midst of the merrymaking a shudder ran down his spine. Suddenly a finger appeared in the smoky darkness of the hall and began to write on the wall:

> 'They drank wine, and praised the gods of gold, and of silver, of brass, of iron, of wood, and of stone. In the same hour came forth fingers of a man's hand and wrote over against the candlestick upon the plaister of the wall of the king's palace: and the king saw the part of the hand that wrote. Then the king's countenance was changed, and his thoughts troubled him, so that the joints of his loins were loosed, and his knees smote one against another. The king cried aloud to bring in the astrologers, the Chaldeans, and the soothsayers . . .
>
> And this is the writing that was written, MENE, MENE, TEKEL, UPHARSIN (numbered, weighed, found wanting). In that night was Belshazzar the king of the Chaldeans slain.' Daniel 5, 4–7 and 25–30

All we can deduce from this event is that the sacred vessels from the temple had magic powers. There is no mention of the ark of the covenant.

Let us take a closer look at Jeremiah. There is something that does not tally.

As his scribe Baruch (11) related, his master was clearly warned by a certain angel of the all-highest *before* the onslaught of the Babylonian army. This angel, who obviously knew what was going to happen, ordered Jeremiah to hide the sacred vessels given to Moses by the Lord from the Babylonians who would attack sooner or later. In other words, the angel was not interested in the vessels which Belshazzar later had carried into the feast, the salvers, chalices and lamps, but in the artefacts entrusted to Moses during the journey through the wilderness. Among these, of course, was the ark of the covenant, together with its manna-making machine!

Jeremiah, who realised the seriousness of the situation, called in strong men, including his *Ethiopian* friend Ebed-melech. Together they carried out a commando raid unnoticed by the people of the city, removed the objects and hid them in a cave. It is a fact that the ark of the covenant did not fall into the hands of the Babylonians, but it did vanish without a trace. No more is said about it in the canonical biblical texts approved by the Church (12).

The only hints we have are in the Apocrypha, the scriptures that were kept secret. The Apocrypha are not accepted into the biblical canon by Christians, although they correspond to the 'official' texts in arrangement and content. The Second Book of Maccabees is such an apocryphical book and in it we read:

> 'And it was also contained in the same writing, that the prophet, being warned of God, commanded the tabernacle and *the ark to go with him*, as he went forth into the mountain, where Moses climbed up and saw the heritage of God. And when Jeremiah came thither, he found an hollow cave, wherein he laid the tabernacle and *the ark*, and altar of incense, and so stopped the door. And some of those that followed him came to mark the way, but they could not find it. Which when Jeremiah perceived, he blamed them saying, *As for that place, it shall be known until the time that God gather his people again together, and receive them unto mercy.*'
>
> II Maccabees 2, 4–7

It says in the *Mishnah** that one of the priests serving in

* The Mishnah is a part of the Talmud and codifies the Oral Law.

the temple went looking for the ark outside Jerusalem one day and found a large boulder which he told his colleagues about, but before he could describe its whereabouts, he died a mysterious death:

'Thus the priests knew that *the ark of the covenant was* hidden there.' Mishnah, Chap. 6, 2 (14)

Once again we have no luck with the ark of the covenant! Searches were not confined to that period, when the whole business was more topical. In 1910 the Parker expedition set out to find it and returned empty-handed.

What had happened to the ark of the covenant?

We are due for a situation report:

—According to the Mishnah, priests suspected that the ark was in the vicinity of Jerusalem, because a priest had died in a mysterious way and his death was ascribed to the ark.

—Tradition indicates that extraterrestrials were active on earth in Jeremiah's day.

—Jeremiah was forewarned by the 'angel of the Lord'. His scribe Baruch writes that there were lights in heaven.

—The prophet Ezekiel's description of his encounters with spaceships also falls into the same period (15).

—Baruch, Jeremiah's friend and scribe, relates in the pseudepigraphical scripture The Rest of the Words of Baruch that the Ethiopian Ebed-melech had an experience with extraterrestrials.

The following assumptions seem logical:

—The group of extraterrestrials was a small one. They did not intervene in battles or help any of the three contending parties. They avoided being seen by large assemblies of people.

—For unknown reasons the group was unable to remove the ark and the manna machine themselves. Were they unwilling to interfere in the affairs of men? Were they, too, afraid of the machine's powerful radioactivity? No matter, for one thing is clear. The extraterrestrials did not want the ark to fall into the hands of the Babylonians. So they asked Jeremiah and a few of his trustworthy companions to hide the hotly disputed object.

—The commando operation meant that others had to be in the secret! Among them was the *Ethiopian* Ebed-melech. Only a short time elapsed between the warning that Jeremiah received and the attack by the Babylonian army. Jeremiah, unable to make a thief-proof hiding place for the ark, was forced to hide it in a cave.

—Given the considerable weight of the ark, Jeremiah and his assistants probably had to use roads or tracks. And as carrying it themselves would have been conspicuous, transport was probably effected by ox-cart. As the operation had to be carried out in *a single* night, the ark must have been hidden near Jerusalem. The Babylonians were advancing rapidly from the west, present-day Jordan.

—Jeremiah appears to have been familiar with the peculiarities of the machine; he probably even knew how to work it.

Nothing happened to any of his helpers! Yet later it killed another priest who came too close to it.

—The extraterrestrials must have known how important the ark was, otherwise they would cheerfully have let the chest fall into the hands of the Babylonians. But they gave orders for it to be concealed.

Where can Jeremiah have hidden or buried this hot potato?

There are countless possible hiding-places in the rugged countryside around Jerusalem. East of Lake Genasereth the hilly landscape is split up by deep clefts and dotted with natural caves—ideal refuges for the ark! Nevertheless, I cannot believe that Jeremiah travelled 130 km with the heavy burden—and 130 km as the crow flies at that! Given contemporary road conditions and the slowness of transport by oxen, it would have taken some days to get as far as Lake

Genasereth. The general direction would have been crazy, too, because they would have run straight into the arms of the enemy.

No matter. Even if Jeremiah did find a hiding-place in the vicinity of Jerusalem, it would be overgrown today. No one has the remotest idea where the divine machine is stored. More important, it is not mentioned again in the historical accounts.

Where *can* the trail lead?

I did not forget that the *Ethiopian* Ebed-melech had witnessed the nocturnal transport of the ark. Could he have talked about the wonderful machine when he returned home?

A criminologist does not give up, even if he has little hope of finding the trail. For a long time I tried hard to lay my hands on written Ethiopian traditions. I knew of the existence of the epic *Kebra Nagast*, which means something like 'Glory of Kings' or 'Renown of Kings'. Hardly anyone in our part of the world has heard of it. It was not easy to get hold of a German translation of the Ethiopian text.

Thank heavens there is one. We have the Royal Bavarian Academy of Sciences to thank for it. They gave the celebrated Assyriologist Carl C. A. Bezold (1859–1922) a stipend so that he could devote his time to translating the unknown work from manuscripts in Berlin, London, Oxford and Paris (16).

We cannot say definitely when *Kebra Nagast* was written, but we should not be far out if we dated the original version to around 850 B.C. Bezold's German translation is based on the text which the two Ethiopians Isaac and Yemharana-Ab translated from Ethiopian into Arabic in A.D. 409. The translators said in the introduction:

> 'We have translated this work from a Coptic book into Arabic . . . in the 409th year of mercy in the land of Ethiopia in the days of Gabra Maskal the king, who is

called Lalibala, in the days of Abba George, the good bishop . . . Pray for me, your humble servant Isaac, and do not blame me for infelicities of expression.'

Of course we must forgive poor Isaac for introducing Christian doctrines and hints about the coming of Christ that were certainly not in the original *Kebra Nagast*. They could not have been, for it was written before Christ was born. How then could King Solomon, who lived about 965–926 *before* Christ, talk about Jesus, the crucifixion and the resurrection?

So it is best to skip the post-Christian additions in order to stay on the pre-Christian trail of the ark of the covenant. In fact we advance considerably in our search, as there is a reference to the ark right at the beginning of *Kebra Nagast*:

'Make an ark of wood that cannot be eaten by worms, and overlay it with pure gold. And thou shalt place therein the Word of the Law, which is the Covenant that I have written with Mine own fingers . . . Now the heavenly and spiritual (original) *within* it is of divers colours, and the work thereof is marvellous, and it resembleth jasper and the sparkling stone, and the topaz, and the hyacinthine stone, and the crystal, and the light, and it catcheth the eye by force, and it astonisheth the mind and stupefieth it with wonder; it was made by the mind of God and *not by the hand of the artificer, man*, but He Himself created it for the habitation of His glory . . . And within it are a Gomor (?)* of gold (containing) *a measure of the manna which came down from heaven*; and the rod of Aaron which sprouted after it had become withered though no one watered it with water, and one had broken it in two places, and it became three rods being (originally) only one rod.'

Kebra Nagast, Chap .17

A plausible description of an apparatus of which the Ethiopians of that day knew nothing. From their

* Possibly *omer*, an ancient Hebrew measure of about 3 litres.

vocabulary they took concepts which at least made matters roughly intelligible. Ezekiel did the same thing when he described the glory of the Lord as resembling sapphire, precious stones and crystal. Enoch made a similar attempt when he depicted the leader of the extraterrestrials in this semi-Surrealist way: 'His body was like unto a sapphire, his face unto a chrysolith . . . a powerful light, not to be described, and in the light were figures . . .' That is what we find in the Apocalypse of Abraham. How closely the images resemble each other!

An important thing about the first mention of the ark in *Kebra Nagast* is the statement that *inside* the ark there was an unusual object which was *nót* made by human hands.

Kebra Nagast gives us a detailed and colourful account. It says that the Ethiopian Queen Makeda learnt from a travelling merchant that Solomon, king of the Israelites, was a handsome man and ruled over a magnificent kingdom. Queen Makeda also heard about the god of the Israelites and the mysterious ark he gave to the wandering people.

This news inspired the queen to pay a neighbourly and friendly visit to her colleague Solomon. She prepared for a sumptuous journey, without counting the cost. It says in *Kebra Nagast* that 797 camels were saddled, countless mules and asses used as pack animals and that there were more than 300 people in her retinue.

According to report, Solomon the Wise, was also an indefatigable playboy—a womaniser of the kind that had no place in the Mosaic law. Not only did he enjoy himself with the women of his own country, he also had playmates imported from over the border. No wonder, then, that he gave the Ethiopian queen a reception that was splendid in the extreme:

'And he paid her great honour and rejoiced and he gave her a habitation in the royal palace near him. And he sent her food for both the evening and the morning meal, each time 15 measures by the kori* of finely ground white meal, cooked with oil and gravy and sauce in abundance, and 30 measures by the kori of crushed white meal wherefrom bread for 350 people was made, with the necessary platters and trays and ten stalled oxen, and five bulls, and 50 sheep, without counting the kids, and deer, and gazelles and fatted fowls, and a vessel of wine containing 60 gerrat measures, and 30 measures of old wine . . . And every day he arrayed her in garments which bewitched the eyes.' *Kebra Nagast*, Chap. 25

The trifles in which Solomon the Wise invested paid off. He seduced the queen right royally and as he spoilt her and gave her so many presents during her stay, he had to be equally lavish when she left. Let us look at the list in *Kebra Nagast*:

'He . . . gave unto her whatsoever she wished for of splendid things and riches, and beautiful apparel . . . and everything on which great store was set in the country of Ethiopia, and camels and wagons 6000 in number, which were laden with beautiful things of the most desirable kind, and wagons wherein loads were carried over the desert . . . *and a vessel wherein one could traverse the air, which Solomon had made by the wisdom God had given unto him.'* *Kebra Nagast*, Chap. 30

This text needs reading twice. It describes in detail the presents Makeda took back to Ethiopia: camels, carts, vessels and vehicles for land travel . . . and a cart which flew through the air! The chronicler distinguishes clearly between the types of cart: one for land travel, one for travels through the air. An astonishing character, this Solomon. He seemed to have everything in his car park!

The inevitable happened.

Nine months and five days after her return, the queen gave birth to a son whom she called Bayna-lehkem. (Now

*Ancient Hebrew measure = 364 litres.

an idea which may be straying from the subject. Phonetically the name Bayna-lehkem is very similar to Ebed-melech. Surely it is conceivable that vowels and consonants got muddled in colloquial speech. That Bayna-lehkem and Ebed-melech are identical? Chronologically this possibility makes no sense, for Solomon's period dates about 400 years further back in the past than the lifetimes of Jeremiah and Ebed-melech. But there would be nothing exceptional about the chroniclers confusing the names when they reported events. But as I have said, this is only by the way.)

Bayna-lehkem, the son of a whirlwind love affair between king and queen, was trained in all the arts and the use of every kind of weapon, as befitted his station. At the age of 22, he, too, travelled to Jerusalem to meet his father:

> 'And the youth Bayna-lehkem was handsome, and his whole body and his members, and the bearing of his shoulders resembled those of King Solomon his father, and his eyes, and his legs, and his whole gait resembled those of Solomon the King.' *Kebra Nagast*, Chap. 32

Delighted by the visit, Solomon lavished magnificent gifts on his son in truly regal manner. But Bayna-lehkem was smart!

None of the wonderful presents excited him. He had one secret imperative wish. *He wanted the ark of the covenant!*

He told his father Solomon so and added that he would like to take the ark to his mother, because whoever owned it was protected by the Almighty.

Solomon was a little frightened when he heard this wish, but only a little, although he knew he would have to keep things quiet. After all the ark was a priceless relic which came from Moses and was kept in a special room in the temple to which only chosen priests had access. In view of the king's attitude, we can assume that, as matters stood, he no longer had any special use for it or that he wanted to protect Makeda by installing the machine in her palace and reminding her of the happy hours they had spent together. In any case he would have entrusted the custody of the ark to someone.

The ark of the covenant was never transported without special precautions.

Solomon made transport conditional on two things:
—that it was to take place in absolute secrecy;
—that it must be done without his official knowledge.

Both conditions were natural. If priests and public found out the king was giving the precious ark away so casually, there would be a revolution.

Bayna-lehkem had inherited wisdom from his father and astuteness from his mother. He consulted his confidants to find out how he could fulfil the parental requirements. They came to the conclusion that a trick was the only solution. As the king's son, Bayna-lehkem was a man of confidence who had access to secret rooms. So he would go into the room containing the ark and take its precise measurements. Then his men would visit the city and order separate parts of the ark from carpenters, without those worthy caftsmen realising what they were making:

> 'And I will take the framework without the pieces of wood thereof being fixed together, and I will have them put together (later). And I will set them down in the habitation of Zion (the ark), and will drape them with the draperies of Zion, and I will take Zion, and will dig a hole in the ground, and will set Zion there, until we journey and take it away with us thither.'
>
> *Kebra Nagast*, Chap. 45

A simple, yet artful plan.

When the carpenters had delivered parts made of the same wood and with the same colour as the original ark, Bayna-lehkem entered the temple room by night, leaving the door ajar so that his friends could follow him. They were to remove the Mosaic ark of the covenant covered with old rags, carry it to the Ethiopian camp outside Jerusalem and bury it, until they set off for home. A copy made of the imitation parts was set up in the room. No one would notice the difference:

> 'And he rose up straightway, and woke up the three men his brethren, and they took the pieces of wood, and went into the house of God—now they found all the doors open, both those that were outside and those that were inside—to the actual place where he found Zion, the Tabernacle of the Law of God; and it was taken away by them forthwith . . . And the four of them carried Zion away, and they brought it into the house of Azaryas, and they went back into the house of God, and they set the pieces of wood where Zion had been, and they covered them with the covering of Zion, and they shut the doors.'
>
> *Kebra Nagast*, Chap.48

A week later the Ethiopians broke camp. No one in Jerusalem had noticed what had happened to the ark—yet another indication that the Israelites were no longer interested in the manna machine, which did not function:

> 'And they bade (the king) farewell and departed. And first of all they set Zion by night upon a wagon together with a mass of worthless stuff, and dirty clothes, and stores of every sort and kind. And all the wagons were loaded, and the masters of the caravan rose up, and the horn was blown, and the city became excited, and the youths shouted loudly.'
>
> *Kebra Nagast*, Chap. 50

Far away from Jerusalem, knowing that they were safe, the Ethiopians unloaded the ark of the covenant and put it on to a new cart. Once again an extraordinary phenomenon took place, but we know that the ark was never transported without something special happening:

> 'Michael the Archangel marched in front . . . and spreading himself out like a cloud over them he hid them from the fiery heat of the sun. And as for the wagon, no man hauled it, but he himself (i.e. Michael) marched with

the wagons and whether it was men, or horses, or mules, or loaded camels, *each was raised above the ground to the height of a cubit*; and all those who rode upon the beasts were lifted up above their backs to the height of one span of a man, and *all the various kinds of baggage* which were loaded on the beasts . . . *were raised up to the height of one span* of a man . . . *And everyone travelled in the wagon like a ship on the sea when the wind bloweth . . . and like an eagle when his body glideth above the wind.* Thus did they travel; there was none in front and none behind, and they were disturbed neither on the right hand nor on the left.'

Kebra Nagast, Chap. 52

A fresh report is due.
—Among Solomon's many gifts to the Ethiopian queen was a car which travelled through the air.
—The king's son secretly removed the ark from Jerusalem, with Solomon's tacit consent. It was hidden outside the city for a week. When he departed it was put on a cart and covered with old rags and rubbish.
—Not until they were far away from Jerusalem was the ark transferred to a *new* cart. This cart flew a cubit or the span of a man above the ground. The cart must have been quite large, for horses, mules and camels were 'raised up', as well as men. The chronicle expressly mentions *one* cart, whereas there were many carts in the retinue on the outward journey.
—The king's son had thought of everything in his plan. He left the big flying cart—obviously the king's gift to his mother—some days' journey from Jerusalem, under a military guard. Then he arrived in Jerusalem like an ordinary traveller, stole the ark, departed and loaded it on the flying cart. The cart could travel fast and was not held up by the miserable roads. Pursuers had no chance of catching up with the caravan.
These assumptions are confirmed in the chronicle.

The temple priests in Jerusalem discovered the theft and told King Solomon. They urged him to assemble troops to follow the Ethiopians without delay. Solomon could not refuse their demands, but neither could he admit that his son had his secret approval to remove the ark.

Even Solomon's fast cavalry could not find the route taken by the Ethiopians. No wonder, for they had *flown* to Egypt and their aerial journey had caused considerable havoc. The Egyptians told the Israelite reconnaissance troops:

> 'And the men of Egypt said unto them, "Some days ago certain men of Ethiopia passed here; and they travelled swiftly *in a wagon, like the angels, and they were swifter than the eagles of the heavens.*"
>
> And those who were in the cities and towns were witnesses that, when these men came into the land of Egypt, our gods and the gods of the king fell down, and were dashed in pieces, and the towers of the idols were likewise broken into fragments.' *Kebra Nagast*, Chaps. 58 and 59.
>
> It is all most remarkable.

A flying wagon that knocks obelisks (towers) over? On which there was room for horses and riders and camels? A product of extravagant oriental imagination?

Enormous flying objects are also described in the Indian epics *Mahabharata* and *Ramayana*. One of them, left behind by the gods, was 'as big as a temple and five storeys high'. The *Ramayana* mentions flying machines 'which (make) the mountains tremble when they take off amid thunder and burn up woods, meadows and the tops of houses'. We should confidently accept the Egyptians' horror story as fact.

Obviously the messengers from King Solomon's blitz troop did just that when they made their report. The king did not give up easily. He put himself at the head of a squad of chosen men and asked the Egyptians when his son Bayna-lehkem had left. This is what he was told:

'He left us three days ago. And having loaded their wagon

none of them travelled on the ground, but *in a wagon that was suspended in the air*; and they were swifter than the eagles that are in the sky, and all their baggage travelled with them in the wagon above the wind. As for us, we thought that *thou* hadst, in thy wisdom, made them to travel in a wagon above the wind. And the King said unto them, "Was Zion, the Tabernacle of the Law of God, with them?" And they said unto him, "We did not see anything".' *Kebra Nagast*, Chap. 58

Solomon realised that his own son had made a fool of him. He understood that he could never recover the ark and its precious contents. Solomon and his priests wept in secret, mourning for the loss of the ark. In secret, because the king realised at once that the theft could not be made public. Now that Israel was no longer in possession of the ark, hostile kings might suddenly feel strong enough to attack Israel. Hence Solomon's strict orders to the priests not to let a word about the loss of the ark leak out to the public:

'And Solomon answered and said unto them, "Cease ye, so that the uncircumcised people may not boast themselves over us, and may not say unto us, 'Their glory is taken away, and God Hath forsaken them.' *Reveal ye not anything else to alien folk.* Let us set up these boards, which are lying here nailed together and *let us cover them over with gold, and let us decorate them after the manner of our Lady Zion (the ark).* And let us lay the Book of the Law inside it."' *Kebra Nagast*, Chap. 62

Solomon was forced to cover up the theft of the ark and to order the priests to decorate the *false* ark with *genuine* symbols. But his end was near. We read in *Kebra Nagast* that he lived for another 11 years, but turned from God and devoted himself to a hectic love life.

What happened to the ark of the covenant once it was in Bayna-lehkem's possession?

When the king's son and his men had flown across the Ethiopian frontier, he gave the order to land, and, like King

David before him, danced for joy around his booty:

> 'And the King rose up and skipped about like a young sheep and like a kid of the goats that hath sucked milk in abundance from his mother, even as his grandfather David before the *Tabernacle of the Law of God*. He smote the ground with his feet, and rejoiced in his heart, and uttered cries of joy with his mouth. And what shall I say of the great joy and gladness that were in the camp of the King of Ethiopia? One man told his neighbour, and they smote the ground with their feet like young bulls, and they clapped their hands together, and marvelled, and stretched out their hands to heaven, and they cast themselves down with their faces to the ground, and they gave thanks unto God in their hearts.' *Kebra Nagast*, Chap. 53

Mother Makeda handed over the throne to her successful son, who was known as King Menelik from then on. He became the founder of the new Ethiopian dynasty.

The Ethiopian constitution of 1955 still says in Article 2 (17):

> 'The royal dignity shall stem from all eternity from the same family line which runs uninterruptedly from the dynasty of King Menelik I, son of the queen of Ethiopia, the Queen of Sheba, and King Solomon of Jerusalem.'

The Negus of Abyssinia, Emperor Haile Selassie of Ethiopia, banished in 1974, traced his dominion back to Menelik. The Ethiopian monarchs sometimes called themselves King, sometimes Emperor, sometime King of Kings, convinced that they were superior to all other rulers and under the protection of almighty God, thanks to the invincible power of the ark of the covenant.

During my search for the trial of the ark of the covenant, I remembered an experience I had in Srinagar, in the highlands of India, in 1976. Professor Hassnain, an archaeologist, took me to visit a mountain called Tahkti Suleiman. A temple, which is a Mohammedan sanctuary, stood on the summit of the steep mountain.

I asked what Tahkti Suleiman meant. Professor Hassnain answered at once: 'Solomon's Mountain!'

I thought it was rather ridiculous to call a mountain in the Indian highlands after a Hebrew king. I asked more questions and was given this explanation by the Professor:

'King Solomon is revered by both Mohammedans and Hindus. This is his mountain and this is the king's temple! It was built here, because tradition has it that King Solomon flew here in a flying machine and personally arranged for its construction.'

At the time this was double Dutch to me. I didn't believe a word of it, but concealed my scepticism because Professor Hassnain is a practising Mohammedan.

Since I have read *Kebra Nagast*, I believe that King Solomon the Wise could have flown all over the world. In the Old Testament Solomon is always reputed for his wisdom; perhaps it should have been for his technological knowledge.

Unfortunately we do not know, nor shall we ever find out what kind of a flying machine he had built. Did the Sons of Heaven, of whom the antediluvian prophet Enoch speaks, leave a shuttle ship behind?

Was there a specially trained priesthood, a secret technical guild, which knew how to work the monster? Big questions which cannot be answered. All we know for sure is that—according to *Kebra Nagast*—Solomon presented the queen of Ethiopia with a *flying machine*, which later played a decisive role in the theft of the ark of the covenant.

The first stage on King Menelik I's flight was the Ethiopian town of Waqerom. Then he flew to the capital, which was called Dabra Makeda:

> 'And the King came with great pomp unto his mother's city, and then he saw *in the height* the heavenly Zion *sending forth light like the sun*. And the Queen . . . threw up her head and gazed into the heavens, and thanked her Creator; and she clapped her hands together, and sent forth shouts of laughter from her mouth and danced on

the ground with her feet; and she adorned her whole body with joy and gladness with the fullest will of her inward mind. And what shall I say of the rejoicing which took place then in the country of Ethiopia, both of man and beast, from the least to the greatest, and of both women and men. And pavilions and tents were placed at the foot of Dabra Makeda on the flat plain by the side of the good water, and they slaughtered 32,000 stalled oxen and bulls. And they set *Zion* (the ark) upon the fortress of Dabra Makeda, and made ready for her 300 guards who wielded swords to watch over the pavilions of Zion.'

Kebra Nagast, Chap. 85

I should mention that in Old Testament commentaries the view is repeatedly put forward that King Solomon was visited by the Queen of Sheba, not the Ethiopian queen. (The kingdom of Sheba was in present-day Yemen.)

It is not absolutely clear from the texts whether the Queen of Sheba also visited Solomon, or whether Queen Makeda was monarch of Sheba as well as Ethiopia (18).

On the other hand it is clear that the ark of the covenant was taken to *present-day* Ethiopia. *Kebra Nagast* gives an exact flight path for the Ethiopians' return. It went from Jerusalem along the coast of the Mediterranean to the Nile, which is mentioned as the 'brook of Egypt'. Menelik's crew used the Nile as a navigational aid, but the Egyptians could not harm the Ethiopians, because they had flown on. Besides the Egyptians knew that the aviators had the dangerous ark with them. This frightened them, because they knew from many reports that the machine was fatal to anyone who tried to take it. It says in *Kebra Nagast* that the ark was as radiant as the sun. Uncanny things went on up there in the flying machine. Were death rays directed at the enemy? Did the flying machine gleam in the rays of the sun? There is no conclusive answer.

The ark of the covenant was not taken to the Yemen across the Red Sea from Upper Egypt or Ethiopia. The boundaries of the Kingdom of Ethiopia were clearly defined:

'And thus the eastern boundary of the kingdom of the King of Ethiopia is the beginning of the city of Gaza in the land of Judah . . . and its boundary is the Lake of Jericho, and it passeth on by the coast of its sea to Leba and Saba . . . and its boundary is the Sea of the Blacks and Naked Men, and goeth up Mount Kebereneyon into the Sea of Darkness, that is to say, the place where the sun setteth . . .'

Kebra Nagast, Chap. 92

The ark travelled through the country. It found its final resting place in the north Ethiopian town of Axum, which was once the capital of the kingdom and is reputed to have been founded by one of Noah's grandsons.

On 24 October, 1970, the *Neue Zürcher Zeitung* reported:

'Nearly 3000 years ago, Menelik I, son of Queen Makeda of Sheba and King Solomon the Wise, is supposed to have taken the *sacred ark of the covenant* from Jerusalem to Axum, *where it presumably still is today in the custody of the priests of the Cathedral of Mary*. Axum owes its possession of this holy relic to its position as the religious centre of Coptic Christianity.' (19)

Are the ark of the covenant and its mysterious manna machine really still in Axum? Axum, situated about 180 km south of the Ethiopian provincial town of Asmara, has become a tourist centre. The tourists can see temples and tombs, and a reservoir, which is called 'the Queen of Sheba's Bath'. They marvel at gigantic steles, the biggest of which stood 33.5 m high before it collapsed. There are supposed to be graves under the steles. But no one knows for sure.

Who can tell if the ark is still in Axum? After the second Italian-Ethiopian war of 1935/36, Ethiopia and Eritrea came under the sway of Rome. Would the Italians have missed the chance of secretly removing the ark to Rome as a trophy? A bold speculation, admittedly. The ark may even be preserved in the Vatican today, thanks to an act of homage by the Fascists. But that is something we do not know, and even if it is true, we shall never find out. It is quite a dangerous idea.

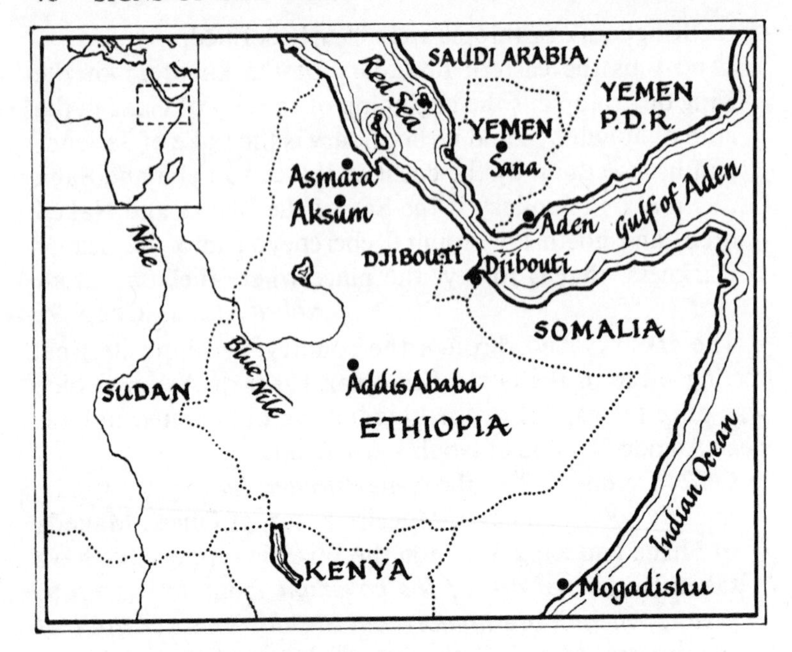

Yet another status report:

Does the theft of the ark of the covenant by the Ethiopians explain why there is no mention of it in literature for centuries? Did the priests obey Solomon's order not to tell anyone about King Menelik's deceit.

A particularly arresting idea is that the prophet Jeremiah hid the *false* ark 400 years after the event! We know that the king's son had a fake chest with *genuine* symbols placed in the chamber, indeed he even put the tablets of the law in the ark for the sake of authenticity. There was nothing to show Jeremiah that it was a fake. He lived about 600 B.C., i.e. nearly 300 years later than Solomon, who died in 933 B.C. So the prophet would have acted in good faith in saving the fake from the Babylonians.

This assumption would contradict my previous speculation that extraterrestrials warned Jeremiah against the Babylonians *because of the ark*. But it is possible that the extraterrestrials had absolutely no idea of its removal to Ethiopia, for there is no evidence of their presence during the period between Moses and Jeremiah and Ezekiel. They had no information about what happened in the interim.

This leaves two possibilities:

1 Jeremiah saved the genuine ark from the Babylonians. If so, it must still be in some cave or grotto near Jerusalem.

2 The genuine ark was taken to Ethiopia by Bayna-lehkem and is hidden somewhere in that country, perhaps in the holy city of Axum.

We still have to clear up the peripheral question of what happened to the flying machine:

> 'But the King . . . and all those who obeyed his word, flew on the wagon without pain or suffering, and without hunger or thirst, and without sweat or exhaustion, and travelled in one day a distance which (usually) took three months to traverse.' *Kebra Nagast*, Chap. 93

King Menelik used the 'flying carpet' when waging war. Clever of him, as none of his enemies had such a machine and he was superior to all of them. I can hear my critics muttering: all right, show us this flying machine! Nothing would remain of it after 3000 years; it would be corroded and overgrown with vegetation. It could also have crashed in other countries, places where no traditions were recorded. May I point out that even today aircraft crash without being found? No, we shall never find the royal gift again. But the chances of getting on the trail of the ark of the covenant are definitely greater.

Why?

We know from the chronicles that the ark and the mysterious apparatus inside it emitted rays, that they were dangerous. The rays must have been powerful, for people who even approached the ark were killed, while others became seriously ill. Effects like that are not caused by reflected sunlight!

The part of the apparatus that produced energy must have

been very small. I have already mentioned a mini-reactor of the kind in use today.

What kind of rays can have been a) as strong and b) as long-lived as the accounts of the ark state? Plutonium would be a possibility. It has a half life of 24,360 years. This means that half the original radioactivity still remains after 24,360 years. The signal, the 'clock' that still ticks today, is as accurate as that!

Present-day technological methods could be used to pinpoint such rays. It is quite simple; all that is needed is a helicopter with sensitive detectors on board to fly over the potential territories. If plutonium was the substance, it would still be radioactive today.

And the ark would still be giving off rays if a different radioactive material was used to power the mini-reactor. The chance may be slender. Admittedly. But it is a chance and today we willingly hand out millions of pounds for projects where the chances of success are much less. Why don't we invest in research into our past for once? We should win the future. If we were successful in our search for the ark of the covenant (and the manna machine), we should at least know that:

1 Extraterrestrial forms of life, superior to us, exist.
2 Extraterrestrials visited the earth.
3 They guided groups of earlier inhabitants of the earth in a specific direction.
4 The oldest rules for the communal life of intelligent beings came from extraterrestrials.
5 We should get to know the ancient technology of the extraterrestrials, their metallurgical skills, in short, their state of knowledge.

I fully realise that there would be religious and political opposition to such a search in Israel, Jordan and/or Ethiopia. You don't have to tell *me* the scepticism that boils up as soon as an apparently Utopian idea is even mentioned. But to be fair, people should not always be asking *me* for *visible proof* of the presence of extraterrestrials, if they are not prepared to make the slightest attempt to follow up the hints that do exist.

The ark of the covenant must still be in existence.

The 'case of the Ark of the Covenant' could have a happy ending even now, if only we wanted it badly enough.

Communiqué

In 1753 the Portuguese Joao da Silva Guimaraes published his *Historical account of a vast hidden uninhabited city of great age discovered in the year 1753*. Today the document is preserved in the state archives of Rio de Janeiro.

Guimaraes tells how he and eighteen companions were looking for gold and diamonds on the River Gonfugy north of the town of Boa Nova. During the months spent in forests and marshes, they had completely lost their sense of direction when they suddenly found themselves on a hill. This moment is described as follows:

> 'Below us lay the buildings of a city surrounded by forest. We passed through a large arched gate on which writing was engraved. We found broad streets, and broken columns lay everywhere. A black column stood in a square and on it was a man, with his left hand on his hip, and his right hand outstretched and pointing northwards. We also visited a hall with many pictures on stone that were badly damaged. There were characters on the obelisks that we could not read. In a ruined hall hung a large disc of rose-red stone . . .'

In 1925 Colonel Percy Harrison Fawcett, a member of the Royal Geographical Society of London, set out on an expedition to find this mysterious city. Fawcett and his companions never came back.

A rescue expedition was organised in 1928. It was unsuccessful.

In 1930 another expedition set out under the leadership of the British journalist Albert de Winton. Winton did not return either.

In 1932 the Swiss trapper Stefan Rattin made a report to the British Consul-General in Rio de Janeiro. He had seen Colonel Fawcett, who was the prisoner of an Indian tribe. From the verbatim report:

> 'Towards sunset on 16 October, 1931, my companions and I were washing our clothes in a tributary of the River Iguassu Ximary, when we suddenly found ourselves surrounded by Indians . . . After sunset an old man wearing skins with a long yellowish-white beard and long hair appeared unannounced. I realised at once that he was a white man . . . He looked very sad and could not take his eyes off me . . . When the Indians were asleep, the old man came over to me and asked if I was English . . . He went on: "I am an English colonel. Go to the British

Consulate and ask them to tell Major Paget that I am held prisoner here."'

Bryan Fawcett, the missing man's son, did not believe Rattin's story of seeing his father and as the son took no action, the Swiss, in a fury, decided to bring the old man back to civilisation on his own account. Stefan Rattin was never seen again.

In 1952 Bryan Fawcett organised his own expedition to look for the father who had vanished from the face of the earth twenty-seven years before. His conclusion after his mission: Colonel Percy Harrison Fawcett and all his companions were murdered by Indians.

And what about the city described in the Portuguese document of 1753? It has never been seen again. No serious official expedition was ever organised. Today the people in the Red Square in Moscow can be counted from a satellite. From a height of x kilometres we can say whether Leonid Brezhnev's *dacha* is heated or not. Airborne sensors can discover minerals and oil deep in the earth from a moderate height.

We can do all kinds of things, but we cannot take the trouble to discover cities hidden in the virgin forest. At least one, the city that Guimaraes came across and Colonel Fawcett looked for, would be a worthy goal. Why does no government or research institute commission a search for this missing city? It would be an excellent project for NASA.

Colonel Fawcett said:

'Whether we get there and come back again, or leave our bones to rot in the interior, one thing is certain: the answer to the puzzle of ancient South America and perhaps of the pre-historic world as a whole may be found when the location of those ancient cities is established and made accessible to scientific research. I know that those cities exist.'

Sources: Bryan Fawcett ed., *Exploration Fawcett*, London, 1953.
Revista, Vol. 1, 1839, p. 181, Document No. 514.

2: Man Outsmarts Nature

It is well known that men make men and that the process is pleasurable. Men made by robots will come in the future. I have claimed that gods made men. I am going to prove that today men are already capable of producing men artificially, like the gods.

• • •

Thanks to the friendly cooperation of Lesley Brown, 32, from Bristol, the world's Press was able to sail through the silly season of summer 1978 with ease.

Mrs Brown was infertile; the oviduct leading to her womb was blocked.

Nevertheless, Dr Patrick C. Steptoe, a gynaecologist, helped Lesley to have the child she longed for. He removed an egg from her ovary and united it in a test-tube—*in vitro*, as doctors say, which means 'in glass' in Latin—with one of her husband's sperm cells. Under the doctor's watchful eye, the embryo flourished in a nutrient solution. At the critical moment he inserted it in Mrs Brown's womb. Louise, born in the summer of 1978, was a healthy child. Just like a baby conceived *in vivo* (inside the body). She differed from other babies in the same year only by the publicity aroused by her birth. No baby ever hit so many headlines or had such detailed commentaries on how it originated. No other baby had so many photographs on the front page or a proposal of marriage in the cradle as little Louise.

The test-tube baby of summer 1978 only became a sensation because parents and doctors were not afraid to publicise the story of its origin. No one can openly state that Brown Jr from Bristol has hundreds and possibly thousands of 'brothers and sisters' of the same provenance, because the SECRET sign is up. They grow up safe and sound in secret, because the doctors who 'produced' them are justifiably afraid of being attacked by scientists and even more so by the Church (although there are voices on both these fronts who cautiously admit that fertilisation *in vitro* is compatible with religious doctrines and ethical requirements).

But look what happened to Daniele Petrucci of Bologna when he announced in the middle of the 50's that he had successfully bred over 500 human embryos in test-tubes and kept them alive. At least one Petrucci child is now alive, of marriageable age and sound in wind and limb. Presumably the young man will not forgo the pleasure of creating his progeny *in vivo*— at least I hope so.

His father, Petrucci, however, forswore further experiments when Pope Pius XII, admittedly without mentioning him by name, issued an unmistakeable warning to anyone interfering with God's handiwork.

As early as the sixteenth century, the Church had to come to terms with the danger of manipulation during the early stages of human life, when Paracelsus, doctor and natural scientist (1443–1541), introduced the then unheard of idea of breeding embryos outside the womb (20). Paracelsus surmised that a *homunculus*, a little man, could be produced, if male sperm kept at body temperature in a vessel was nourished with essence of human blood.

Paracelsus's bold vision inspired Goethe, in *Faust Part II*, to include a homunculus produced in a laboratory according to this recipe. His helper Wagner was delighted with the result:

'A human being in the making!

. . .

A mighty project may at first seem mad,
But now we laugh, the ways of chance foreseeing:
A thinker, then, in mind's deep wonder clad,

May give at last a thinking brain its being.

. . .

Now chimes the glass, a note of sweetest strength,
It clouds, it clears, my utmost hope it proves,
For there my longing eyes behold at length
A dapper form, that lives and breathes and moves.
My mannikin! What can the world ask more?'

300 years later all that is left of the alchemist's kitchen is the fact that in ultramodern laboratories scientists are also experimenting in the strictest secrecy and removing the aura of the miraculous from many miracles.

Manipulation of hereditary factors has become possible with the headlong advance of genetic molecular biology since the middle of our century. Here we are talking particularly about molecular genetics, which is concerned with the molecular bases of hereditary, mutation, the exchange of hereditary systems, etc. In other words, it investigates the secrets of the cells of which all organisms are made up.

To get even a remote idea of *how* difficult research into this 'microcosm' is, you must realise that a man has about 50 billion cells in his body. To give only a few comparative sizes, the sperm cell is 0.05 mm long, the largest, the egg cell, has a diameter of 0.1 mm, whereas the nerve cell has a diameter of only 0.008 mm. Nevertheless, the secret code, the building plan (DNA), according to which the *whole* plant, the *whole* animal, the *whole* man, is formed, is in every cell. One cell being 'born' from its predecessor is a very logical way of building on nature's part. To put it as simply as possible: if only one of the 50 billion cells stays alive, the *whole* man could be reconstructed from it. To put it another way, it is as if every stone of St Paul's Cathedral was impregnated with the ground plan and façades of the whole building.

You would imagine that Dr Steptoe received unanimous congratulations on his success in creation *in vitro*. How wrong you would be! He did not suffer as much as his

American colleague L. B. Shettles, who, at the insistent request of a couple from Florida, also succeeded in fertilisation *in vitro*, but was hounded from his university before implantation took place (21). But even Dr Steptoe was accused of 'degrading mankind' and labelled 'immoral'. Abuse poured in from all sides.

I do not understand what is immoral about doctors helping married couples who want babies to achieve happiness! But there exists the 'Order of Pessimists' (22), which resists every advance and raises a hue and cry condemning every kind of technological and scientific success, whether it be the peaceful use of nuclear energy, or a step towards interstellar spacetravel. In the free-speaking west at least, the 'Order of Pessimists' has a broad field with a free range of fire and they stand ready to attack outside every research station.

These destroyers of the future must be obsessed by gloomy thoughts. Obviously they can only imagine the fruits of research being used in a negative way. Progress is equated with the annihilation of mankind and Armageddon. I prize reason and human responsibility higher than the professional pessimists do. *We* shall remain masters of anything the human mind may produce, just as we have done during the thousands of years of the history of civilisation. For all time.

The replication of men *without* natural fertilisation will be achievable in the forseeable future and be far more successful than the implantation of test-tube babies in the womb.

What I predicted ten years ago—to tell the truth, I am amazed I was so daring—has since come true *in vitro*.

At the time I read:

'In the day that God created man, in the likeness of God made he him; Male and female created he them; and blessed them, and called their name Adam, in the day when they were created.' Genesis 5, 1–2

And:

'And the LORD God caused a deep sleep to fall upon Adam, and he slept: and he took one of his ribs, and closed up the flesh instead thereof; And the rib, which the LORD God had taken from man, made he a woman, and brought her unto the man.

And Adam said, This is now bone of my bones (!), and flesh of my flesh (!): she shall be called Woman, because she was taken out of man.' Genesis 2, 21–23

At the time I asked whether humanly intelligent beings could not have been programmed by an artificial mutation of the genetic code and I also questioned whether our charming mother Eve might not have originated without mating by the removal of a male sperm cell.

It is conceivable. The Sumerian cuneiform character for rib is 'ti' and that means 'vital power'. Should a modern translation of the Bible read: 'God took vital power from Adam'?! But vital power is the *cell*. Without it there is no life, not even in paradise.

Today research into molecular biology is based on this fact.

It is easy to be clever with hindsight.

My former questions did not probe deep enough. I should also have asked how Adam came on to the scene out of the blue. Which was there first: the egg, the cock or the hen? Adam *may* have been a test-tube baby, but he may also have been produced as a *clone*. I am interested in this first clone and now I want to set down some ideas on the subject which I hope are not too audacious.

To which race did the test-tube baby made in Oldham belong? Naturally to the white race, because its parents were white.

But to which race did our ancestors—let's call them Adam and Eve—belong? Were they white, black or yellow? Did they have skins with other colours, which are not found today?

The evolutionists say that man descends from monkeys. Yet who has ever seen a white monkey? Or a dark ape with curly hair such as the black race has?

No one will deny that our physical structure indicates a relationship with the apes, that there are similarities, such as using the hands as prehensile tools and the large eyes facing forwards which facilitate spatial vision.

All this is admitted, but there must have been an additional element. In my view an extraterrestrial one. A cross between the planet earth (animal) and the cosmos (intelligence), because intelligent man as he exists cannot simply have descended from some pre-ape species. His race alone proves that.

So which race did the first man belong to?

Why are there different races anyway?

Ethnology is a branch of biological anthropology and the history of human races. By race we mean the subgroup *of a species* which differs from another subgroup of the *same* species by varies external traits. Such traits may be proportions, shape of face, colour of skin (caused by pigment, a material with its own colour appearing in the cell), hair, position and colour of eyes, shape of lips, blood groups, etc.

According to the 1951 definition of race by UNESCO, the three major races—Caucasian, Mongoloid and Negroid—differ from one another by having their own well-marked characteristics which are mainly determined by heredity.

All human racial groups are part of a species, that is to say the three major races with their splinter groups all over the world belong to a single biological species. Species are populations whose individual members can breed with one another. They are able to mate because there are no physiological or morphological limits on human races. This fact is 'proved' daily all over the world . . .

But this does nothing to explain how the different races

originated. There are many theories, but no firm scientific confirmation that it happened in such and such a way and no other! It is certain that races did not originate during the historical period known to us; they have existed since the earliest times.

The ancient civilisations of Sumer, Babylon and Egypt treated racial differences as if they had always existed. Herodotus, Hippocrates and Aristotle mention different races as a matter of course. Racial polemics, accounts of racial wars and hideous pogroms run like a blood-red thread through the millennia of oral and written tradition. One race has always felt superior to the rest at some period or other; members of one race felt provoked by the representatives of another race simply because they were different.

In our century confusing the representatives of *one species* culminated in Hitler's racial lunacy. The resulting trail of blood and murder will live as the prototype of inferno for the rest of human history. As the writing on the wall for all future generations. Present-day communications between countries, peoples, races and tribes will make the feeling and knowledge that we are all members of *a single species* part of the general consciousness. At least I hope so.

This having been clearly and unmistakably stated, there remains the question: why are there human differences? Human genetics, a modern branch of racial research, is trying to make an objective classification enabling us to pick out *genetic* characteristics. Blood groups, serum proteins and Rhesus factors are examined to find classifying characteristics and compared in tabular form as representative of the various races.

Thus it was discovered that 89.3 per cent of all Indians belong to blood group 0, whereas only 0.8 per cent of the Indian population has blood group B. Results for the Mongoloids are different: only 18.3 per cent belong to blood group B and 55.7 per cent to blood group O.

Such comparisons of blood groups are no doubt very interesting to the human geneticist, but I ask myself to what significant conclusion they can lead. For the classifications just mentioned are only valid for the present! How can they tell us whether blood groups have not changed down the millennia or what they will be in the future?

Moreover I can see this method providing new motives for racial conflict. Once a Yankee used to say: 'He's only a nigger!', a negro: 'He's only an Indian!' In future people will say just as pigheadedly: 'He's only a blood group A Rhesus factor + (positive)!' And if it should turn out that one particular genetic combination *is* superior to another, we shall be in the midst of a new *scientific* racial polemic.

But whether they use external or genetic characteristics, racial comparisons will not answer either of my questions: to which race did the first man belong, and why are the qualities of the three major races so fundamentally different.

The negroids (very clearly recognisable among the inhabitants of Jamaica) have dark skins, protruding lips, (predominantly) curly hair and broad noses. (They share few characteristics with the Caucasians.) Within the major negroid race there are 18 subgroups with very different characteristics. So far the Mongoloids have had to be subdivided 20 times. The reason is clear and simple: in the course of the history of evolution deviations from the pattern of the major race were caused by mutations—alterations of hereditary factors. But I am not concerned with comparisons *within* the major races, but only with solving the problem of how the first major races originated.

The starting point for consideration of the subject is the common possession by all races of the same anatomical physical structure and the fact that all races can mate. *All* members of *all* races have the *same protein structure* in their cells. In this connection we once again find an affinity with our simian ancestors—chimpanzees have the same protein structure as we have!

How can that be?

Since Charles Darwin (1809–1892), it is accepted that the species developed by natural selection and that ape and man developed separately from a single product of nature as from a point x in time. In an evolutionary process that lasted millions of years. It may be so. In other words, millionfold mutations over millions of years have made us the crown of creation. Sounds good.

But we must invoke the aid of miracles if people are supposed to believe that hundreds of essential differences were formed during the non-stop series of mutations, whereas the protein structure of chimpanzees and men survived the ennobling process unchanged.

The 146 macromolecules in the protein of haemoglobin (the colouring matter of the red corpuscles) actually are identical in chimpanzees and men down to a single aminoacid building stone. Given so many similarities, we must forgive the Swedish botanist Carl von Linné (1707–1778) for calling the chimpanzee *homo troglodytes*, i.e. caveman.

This identical protein structure in man and chimpanzee proves that man *cannot* have originated *solely* through natural mutation and evolution. Why not?

If we compare the protein structure of two frogs, we find variations 50 times greater than those between chimpanzee and man—yet one frog looks the spitting image of another. Conclusion: as frog and frogs are *more closely* related than man and chimpanzee, the protein structures of the frogs ought to be virtually the same and those of man and chimpanzee totally different. The opposite has been proved.

When Professor Alan C. Wilson and his colleague Mary Claire King, both biochemists in the University of California, saw the astonishing results of their protein investigations, they were convinced that there must be a *hitherto undiscovered and far more effective evolutionary driving force* (23) than anything known so far.

What can this force have been? Professor Loren Eiseley,

the anthropologist fron the University of Pennsylvania, has already said loud and clear that a factor which produced mental abilities during the formation of human groups must have been overlooked by the theoreticians of evolution. That is my opinion, too, but how can we explain the phenomenon that man and chimpanzee are (are supposed to be) more closely related than the popeyed frogs whose protein structures are so different? And this although—according to Darwin and his followers—the period of evolution from chimpanzee to man is supposed to embrace more millions of years and more millions of mutations than the relatively short leap of the frog through world history allowed.

My answer:

There was an artificial mutation from ape to man. We did *not* separate from monkeys so many millions of years ago as is claimed—the family break-up took place only a few tens of thousands of years ago. *That is why* the protein structures of men and chimpanzees remained the same. If millions of years and many thousandfold successful mutations lay between the primitive hominid, a man-like ape, and *homo sapiens*, then—any geneticist would confirm this—the protein structures of both beings would have taken a very different form. Corollary, as this is not so, as they are absolutely identical, our ancestor, the first *homo sapiens*, can only have separated from the ape tribe 'recently', a few tens of thousands of years ago.

Men cannot breed with apes, because intelligent man undoubtedly forms a species absolutely different from any species of ape. How could the human species have developed in such an incredibly positive way within 'minutes', reckoned on the time-scale of universal history? How did ape—or man?—suddenly lose his fur? How did he suddenly think of 'civilising' himself, of creating cultures? Who gave him the idea of hunting animals, whose companion he had only recently been? Where did he suddenly get the illuminating idea of making fire to cook his soup on?

Yes, and with *whom* did the first man mate, when he, a solitary being mutated from the ape tribe, had no suitable sexual partners? He could not mate with his monkey ancestors, for they had a different chromosome count.

Ridiculous nonsense, I hear the anthropologists say. All those things did not happen 'so suddenly'; they took millions of years of gradual evolution. The 'suddenness' of *homo sapiens's* becoming intelligent is an invention of mine with absolutely nothing to justify it. That objection does not hold up, since it is fully established that man and chimpanzee both have the same extremely complicated protein structures.

Where must we look for the eagerly sought but so far undiscovered evolutionary driving force (Wilson)? What is the factor—missed by the theoreticians of evolution—which gave the first human groups mental abilities (Eiseley)?

All these questions are answered as soon as we have the courage to think the apparently unthinkable.

Extraterrestrials separated *homo sapiens* from the ape tribe and made him intelligent by artificial mutation. In their own image. The evolutionary driving force is to be found in this *deliberate* manipulation. It worked perfectly, as we shall see.

I base the following speculative ideas on their successful intervention.

Which race did the first men belong to?

Undoubtedly the structure of the human body derived from a species of ape. So the first men ought to have been black, of negroid race, like their ape relatives. If that was so, why didn't the first 'owners of the earth' spread out over the whole planet? And where did the Mongoloids and Caucasoids, the 'yellows' and 'whites', come from?

Were the extraterrestrials able to opt between different races from the beginning? Did they endow different human groups with different abilities to survive in different climatic and geographical conditions? Was the pigmentation of dark skin genetically programmed so that the race could settle in hot zones? Vice versa, what advantages would white skin have had? Would it have been confined to more temperate zones?

Today it is assumed that primitive men had dark skins. Then the colour of men's skins changed and assumed different shades depending on the time spent in different parts of the world and the amount of ultraviolet rays encountered. Even if people want to make the vitamin D produced by ultraviolet rays responsible for this, it seems to me rather a feeble theory, for Eskimos, who live in icy wastes with little sun, are dark-skinned and surely you are not going to tell me that they get their colouring from fish blubber. Again why are the Mongoloids yellow? And should not black people acquire light-coloured skins to enable them to survive in regions with little sun?

It is possible that the extraterrestrials, with their highly advanced and superior intelligence, deliberately produced different basic races, because they knew from their reconnaissance of our blue planet the different environmental influences to which their creatures would be exposed. In mutating hominids 'in their own image' and making them intelligent, they laid trails for future generations—hints as to their former presence.

As I also attribute high ethical responsibility to a high intelligence, the genetic introduction of differently coloured skins (and other characteristics) may have been meant to have a powerful educative effect. Look around you. No matter what the colour of your skin, you belong to the same species, so live in peace with one another!

Did the crew of the first prehistoric spaceship already belong to extraterrestrial races?

Did they lie with the daughters of earth and produce children, as the great legends of human history tell us? Did this sexual traffic, which was against the orders of the 'gods', originate different races following the model, the genetic pattern, of the extraterrestrials?

Were there, I ask myself, visits at *various points in time* by spaceships which had *no contact* with each other? Did an original group separate *homo sapiens* from the ape tribe and so leave a black race behind? Did another cosmic expedition take place thousands of years later with a white or yellow crew? Was the black race a failure and did the extraterrestrials change the genetic code by gene surgery and then programme a white or a yellow race?

Racial theorists will file my reflections away in the archives. They are satisfied with the current accepted explanations, but what do they really know?

One example will illustrate *how* worthless our previous knowledge is:

A black family emigrates from its home in the tropical zone of the earth and settles in a cooler region. Pigments change down the generations, dark skins become light, perhaps so light, the negroids become white. Dark skin, say the racial specialists, no longer being necessary as a protection against the sun. OK, but in his new environment the black man would also have to lose his curly hair, his prominent dark eyes and protruding lips, otherwise he could never become a white man.

But it's all quite simple, someone will tell me. The black breeds with a white and there you are . . .

Just a minute! I am talking of the time when there was only one race! For in the beginning, and there I am in complete agreement with the racial theorists, there was only the black race, which took its colour from the apes.

But the change from black to white could not have been made with *one* mutation; it would have needed endless chains of mutations.

How does a new species appear, when only one is in existence? How could a washproof black become a white without inter-breeding between two races?

Present-day 'intermediate races', Arabs, Eskimos and

South Sea Islanders, for example, originated by racial inter-breeding. I admit that.

But this possibility did not exist in the very beginning. Science says that only one race existed then and it is supposed to have changed into another race on its own! Or into several races!

We are in agreement on this point: it is science that says that in the beginning there was only *one* race, the black one. Neither a white nor a half-breed was available for inter-breeding. Zero. There were only blacks. I have got that into my head.

The white race, I conclude, cannot have been produced down the generations by blacks inter-breeding with whites. According to that theory, we should only exist if the blacks had mated exclusively with whites for x thousands of years. The possibility exists, but where do we get the whites from?

The Arabs tell a revealing story about the origin of their race:

> 'One fine day the good Lord took a lump of clay and fashioned the first man out of it. He puts his handiwork into the kiln to make it firm and lasting. Rain came down and put the fire out prematurely, and when the good Lord opened the kiln his creation looked white and unappetising. So as not to waste anything, he took out his work, blew life into it and let the white man go forth, in spite of his poor quality. Once again the Lord pulled clay out of the trough to make the second man. He fired the kiln, waited until it was very hot and put his second attempt in. He joined the other gods at a gay dance and forgot his handiwork. When he finally took him out, his second man was black and rather unattractive, but he breathed his life-giving breath into him, too, and sent him forth, because he was not too pleased with him. Then the good Lord decided to create his masterpiece, a man more handsome than either the black or the white. Once again he pushed the well-shaped clay into the kiln, into the

comfortable heat and waited for the right moment to put the fire out. When his man had acquired an appetising brown colour, he took him out and brought him to life with his breath, and as he seemed such a good specimen, he made a pact of friendship with him. That is how the brown man, the Arab, originated.'

This legend makes it easy to guess that the Arabs look on themselves as the chosen race. This arrogance is not confined to the Arabs; it is an attitude unfortunately still shared by members of other races.

There's no better way of burning your tongue than speaking about the white-hot theme of race, not mention burning your fingers if you write about it. There are many reasons why that is so. Apparently it is not only external appearances which make it impossible for the glowing iron to be cooled off in peace and rational calm. What goes on in the heads and 'hearts' of members of another race? The different psyches with their mutually incompatible attitudes make understanding so difficult. The European shakes his head hopelessly when he sees coloured people on television accompanying a dead man to his last rest with loud tomtoms—a situation in which 'one' should behave quietly, ceremonially and sadly. We know from tales by the great eastern story-tellers that they accept with stoic calm blows of fate which would make us reel. Racial characteristics are only external appearances, in the truest sense of the world. The real barriers lie deeper. To penetrate them, we must first concern ourselves with the measurable, recognisable, organic peculiarities. Only then will the deeper 'soundings' be possible that will free us from our ultimate prejudices (and superiority complexes).

Race researchers are still snaking their way through a slalom of flags with questionmarks on them. I should like to know if the there are races with specific qualities which make them capable of special achievements *eo ipso*. Nearly all negroes are musical; they have rhythm in their blood.

Why? Is it only their skin which makes Tibetan sherpas less sensitive than white men to the sun's rays at high altitudes? Why does a black man stand burning sun better than a Mongoloid? Why does no hair grow on the chests of genuine male South Sea Islanders? Why does the hair of the descendants of the Mayas, who live in present-day Central America, never go grey, even in advanced old age? Why do blacks never have blue eyes? Are there races which are obviously endowed with higher intelligence than others?

A list of such questions could easily fill an urban telephone directory.

I quite understand that I am playing with dynamite if I ask whether the extraterrestrials 'allotted' specific tasks to the basic races from the very beginning, i.e. programmed them with special abilities.

I am not a racialist. I do not ascribe advantages or disadvantages to any race on earth. Yet my thirst for knowledge enables me to ignore the taboo on asking racial questions simply because it is untimely and dangerous. In my opinion, black, yellow and white race researchers should combine to dig deeply into the question; why are we like we are?

Once this basic question is accepted, we cannot and should not avoid the explosive sequel: is there a chosen race?

If we take the Bible as the breviary of western wisdom, i.e. if we follow the text of the Old Testament, the Jews consider themselves the 'chosen people'. Should we not ask: chosen by whom for what purpose? Were they singled out for a special task? Is their thousand-year-old claim to be the chosen people one of the reasons for the frightful recurrent persecutions of the Jewish people? Do other peoples feel wary of this claim, do other races think they must defend themselves because of it? Why? The Jews have never done *them* any harm.

When I take a look at the history of the natural sciences in the 19th and 20th centuries, I see that more than half of all scientific achievements and discoveries were made by members of the Jewish *people*. Jews have always held leading positions among astronomers, biochemists, mathematicians, botanists, physicists, doctors, zoologists and biologists (24). From 1901 to 1975 there were 66 (!) Jewish Nobel Prize winners.

So are the 'chosen people' also a 'chosen race'? Certainly not, for the Jews are not a 'race' at all. The great majority of them belong biologically—like their Arab neighbours—to a subgroup of orientaloids of the European race! Hence we do not speak of the Jewish race, but of the Jewish *people*. Thus the outstanding achievements of Jewish scientists cannot be connected with racial aspects.

Nevertheless, whether it fits into the contemporary scene or not, and even if it is unpleasant to sensitive ears, I claim that extraterrestrials did choose a specific race. Mythologies relate how certain 'gods' guided their own 'race', protecting it from hostile alien influences and placing its members in leading positions on our planet. The ancient sources do not tell us which race had special divine advantages, but we find many indications in the Old Testament that the elect were not to mix with others.

When Moses led the Israelites out of Egypt through the wilderness to the promised land on a trek that lasted 40 years, he forbade them any contact with other races, on God's orders. And God watched over and was with the Israelites; he led them and accompanied them with a sign. Ahead of them went a pillar of cloud, which was white by day and shone like fire by night. In this way their jealous God protected them from enemies and strangers, and he fed them with manna, the miraculous bread.

At the end of their 40 years' journey, the Israelites entered their home, the promised land, but only the *new* generation was allowed in. Access to the land where milk and honey flowed was strictly forbidden to the old people, including Moses.

What had happened?

I do not feel that the absurdity of this order can be got round by theological and historical interpretations.

I do not know if the idea I put forward ten years ago has become old hat or a smart up-to-the minute theory. Anyway, my reason for the quarantining of the chosen was that the 'gods' or extraterrestrials—which comes to the same thing—had formed a new generation with new genetic qualities during the 40 years' journey through the wilderness, qualities which the men in the surrounding world did not possess. Was this compulsory isolation of the new genetic material the origin of the still valid rule that Jews should only mate with Jews? Has sticking to this Mosaic maxim preserved not a Jewish race, but a special 'species' of men, who exhibit special advantages and disadvantages vis-à-vis the rest of mankind?

Precisely in our own time, when racial prejudices are on the increase, such remarks may seem inopportune, because a whiff of racism may hover over them. I am aware of the responsibility inherent in raising the question of a chosen race, but I do not think that hushing up problems can ever help to solve them.

Inter alia, modern human genetical research produces racial specifications. In other words, it is treading on the same hot tin roof. One day it will undoubtedly tell us which genetic combinations of a race or species are beneficial and which should be eliminated. To put it in more concrete terms: if a defect in our DNA suddenly meant that we were all born with three fingers and one ear, everyone would be very glad if the fault could be cleared up quickly. Interference with the natural 'patterns' of the cells can be compared with what happens in the vegetable kingdom when a more resistant kind of shortstemmed corn is achieved by experiment, i.e. by programming the cells differently, or in the animal kingdom, when cows with a high milk yield are 'developed'.

Will manipulation of man's hereditary factors be possible in the forseeable future? This is a frightening spectre that looms ahead of us.

What are we to say about this matter of fact text?

'John Gurdon, the Cambridge University biologist, took embryo cells from a female albino frog. He removed their cell nuclei with the hereditary layout and put them into the egg cells of another female frog, from which the cell nucleus had previously been removed. Tadpoles developed from these egg cells. They turned into albino frogs, but they are not related to their mother.' (25)

The procedure used in this experiment was called *cloning*, derived from the Greek *clon* = branch. Günther Speicher makes it easily understandable:

'The cutting of a plant that becomes a new plant when put in the earth is a carbon copy of the mother plant.' (25)

We must always remember that *every* organism consists of cells, each of which contains all the information needed for the reconstruction of the whole organism.

From this microbiologists and microsurgeons concluded that it must be possible to reconstruct the whole from a single cell (without fertilisation) so long as it was possible to remove the nucleus from a cell and then implant it intact in a cell with its egg removed. Once that was done, scientists suspected that it would be possible to multiply every animal, vegetable and human organism exactly after the pattern of the nucleus of the donor cell. It would be impossible to tell the new product from the original. The game that nature sometimes plays with one-egg twins, who are as alike as two peas, could be repeated artificially and without limit.

Professor Gurdon followed this method to the letter when producing his colony of frogs, and every frog resembled the other as closely as frogs do to our superficial gaze. But in this case each of the numerous frogs was a 'genuine' copy of the original. With no mistakes.

Mice are mammals and the first cloned mice are alive!

Test-tube mice. Shortly after a mouse egg had been fertilised *in vitro*, a hair-thin cannula was used to remove the male cell nucleus from the egg cell. Consequently the mouse embryo no longer had the hereditary information of both parents, but only that of the mother, an exact copy of whom it would become.

In other words this method makes it possible to clone females exclusively. What luck! So what about cloning men? It's quite simple! Listen to Professor Illmensee of the University of Geneva:

> 'If we exchange the total hereditary material of a fertilised egg cell for the nucleus of a body cell, we can obviously make copies of male individuals as well.' (25)

The English physiologist Alan S. Parker was almost prophetic when he thought it possible to isolate a human cell nucleus and transfer it to a uterus, long before the possibility of cloning had been repeatedly proved experimentally. He even went a step further when he called for more intensive research into how long male sperm could be stored. He was obviously thinking of the replication of valuable material. Parker was in a very good company, for Professor Marshall W. Nierenberg, who was an important collaborator in the discovery of the genetic code, also thought that every difficulty would be overcome one day. The only question was: when. He conjectured that cells could be programmed with genetic information within the next 25 years. Professor Joshua Lederberg, a geneticist at Stanford University, California, shares his optimism. He is convinced that hereditary factors will be manipulable during this millennium.

It looks as if the experts have been too cautious in their estimates. Everything will happen much faster than we think.

Are we beginning to play the role of fate?

Can't we help acting as we do?

Are we living inside an armour plating of mental processes which we have to follow, because they are programmed in us? Because those who created us made us intelligent 'in their own image'?

Because they were aware that one fine day we would *repeat* what they themselves had tried out on us?

Did not the gods predict this in Genesis:

> . . . this they begin to do: and now nothing will be restrained from them, which they have imagined to do.'
> Genesis 11, 6

Will it become possible to clone men one day? To produce images of a type in any desired quantity? Successful experiments on mammals are (nearly) always repeated on men after a certain lapse of time. Whether the first man cloned by the process minutely described and fully documented by the science journalist David Rorvik (21) is still alive in the USA without anyone knowing, as Rorvik claims, is a question of secondary importance. Beyond the knowledgeable description of an individual case is the fundamental recognition that it will be possible to clone men in the easily forseeable future.

There is always a purpose and target to stimulate every piece of research. How then can the horrific prospect of a multiplied man taken from one cell—whether from the blood, skin or some other organ—be meaningful and useful?

Once the method is practicable, shall we mass-produce politicians, soldiers, scientists, space pilots, workers, priests, soothsayers and comedians? Will the abyss that Orwell and Huxley pointed to open before us? Shall we create new 'racial categories' which will then fight each other because of their peculiarities? Shall we let past fashionable ideals of beauty lapse and use cloning to have male and female mannequins roll off the assembly line? Shall we create types of men specially suited for particular research purposes? Will a man or a woman hoard a couple of cells from his or her beloved spouse to have the original recreated in the case of sudden premature death?

Will tiny remnants of the cells of intellectual giants, geniuses in every field, be on call in 'cell banks', so that a new but identical man in their image can tread in the footsteps of the deceased?

I feel that a great chance for humanity would open up if the knowledge of the genius of the century were not lost with his death. How would the course of the world have been determined if Einstein had been virtually immortal? By the cloning process. The great thinker left instructions for his body to be cremated and his brain to be donated to research (27). It is shameful to learn that this bequest to science is in a glass jar full of formaldehyde in a cardboard carton in the office of a biological experimental laboratory in Wichita, Kansas. Parts of the brain went to specialists; cerebellum and sectors of the cerebral cortex were not dissected. Formaldehyde has a strong germicidal effect. It is most unlikely that a single cell has survived the 34 years since Einstein's death.

No one can guess whether the great savant planned more than a purely academic examination. Did he forsee possibilities that no one could imagine in 1955? Has science destroyed a fantastic possibility for x day?

My speculation, but one motivated by the current state of cellular biology and microsurgery, is that extraterrestrials created *homo sapiens* by cloning, which they already knew all about. If they were masters of interstellar spacetravel, with outstanding technological knowhow, we can well believe that they were experts in genetic manipulation. They 'planted' the DNA of their race and transmitted it intact. From then on the 'divine' programme for the building up of man pursued its course. We are hunting for this primordial knowledge; we carry it within us; all we have to do is rediscover it.

In the decades to come we shall break through into interstellar space. The step will *have* to be taken, because supplies of raw materials on our planet are running out. This need will be a stronger stimulus than man's curiosity to discover unknown peoples or even civilisations in the universe.

It makes no difference what spurs us on; we shall have to penetrate the cosmos for our own survival.

If an uninhabited planet like earth is found in the depths of space, it will be only logical to want to colonise it. Previously thare was one prickly point in the array of arguments against space travel and similar projects. We could hardly transport hundreds and thousands of men and women to the goal in gigantic space ferries, for the cost would be prohibitive and the advantages questionable. Besides even if there was a planet anything like earth, our colonists would only settle there under protest. Combinations of gases insupportable to us and different bacteria would harm our 'race'. Under such conditions, how would our colonists ever get acclimatised? Perhaps there would also be differences in temperature varying from minus 80° to plus 80° centigrade on the hypothetical planet. How would men withstand them without heavy protective clothing (which would hinder them in any kind of physical work)?

Discussions about this and other points carried out behind tightly closed doors came to one conclusion. Cloning! if the planet were uninhabited, a race suited to the conditions of the new planet would be programmed. If there were unintelligent life, human hereditary factors would be introduced into the egg cells of the most developed species. History repeats itself. We shall be doing what the extraterrestrials did to primates on our blue planet!

Does the earth hold any indications or points of reference to support my audacious ideas?

—Many mythologies and the traditions of ancient religions say that the 'gods' created men in their own image and that they had to make several attempts before they were successful (28).

—Several peoples claim, some of them even today, that the rulers of their dynasties are direct descendants of the 'gods'. . . for example, the Egyptian Pharaohs, the ancient Sumerian kings, the Ethiopian and Persian royal houses, the Japanese imperial house, etc.

—The Toradja, a South-Sea tribe in the Sulu Sea, swear that they came from heaven and that their ancestors, the Puangs ,originally had white blood in their veins, until it turned red through mixed marriages with earth dwellers (29).

—Until 1962 the Uro tribe lived on reed islands in Lake Titicaca. The Uros had black blood. They did not mate with the members of neighbouring Indian tribes, because they were convinced that they came from space and wanted to preserve their exclusive origin. They lived a solitary and withdrawn life, always on the move in order to avoid contact with other tribes. Originally the Uros lived on the shores of Lake Titicaca. Not until the warlike Indian Aimara, over 1400 years ago, and later the hordes of the Spanish conquistador Francisco Pizarro (1478–1541) stormed the Bolivian plateau did the Uros build the reed islands on which they lived from then on. They were disdainful of other tribes, but avoided any kind of conflict. Their special qualities led them to a certain arrogance. They said that they did not die in water or feel icy cold. That violent storms could not affect them, that the damp cold fog which made the other Indians ill did not bother them, just as the 'fire from heaven' (lightning) did not harm them. The Uros conversed in a language unknown on this earth.

They obstinately preserved the belief that they were not men. In 1960 there were still eight genuine Uros on the reed islands of Lake Titicaca. The last of them died in 1962.

What race did these conceited hermits belong to? From the beginning of their existence they did not defile themselves by breeding with terrestrials, so they may have preserved a race which remained unchanged from its creation to its demise. Who created these Uros and for what purpose? Were they destined for a special mission and did they fail to carry it out?

If the basic races of mankind are in some way connected with 'my' extraterrestrials, we must ask whether the 'gods' wanted a mixture of race or strict segregation.

If we look for the answer in legends, myths and early religious traditions, the jealous gods were opposed to a racial pool. To avoid repeating what I have long been saying and writing on the subject, I shall only remind you of the quarantinelike separation of the new generation that grew up during the 40 years in the wilderness from the older one, the strict isolation of the Puangs and Uros, and that the ancient Egyptian Pharaohs regularly committed incest to keep things 'in the family'.

We know that all races of the same species can interbreed. If the 'extraterrestrials' did not want this racial miscegenation to take place, they could have imposed genetic limitations by providing sexual organs which were unsuitable for intercourse with other races or by changing the chromosome counts. In other words, the unvarying human chromosome count is the secret code for intelligence! Is that why *every* intelligent being has 46 chromosomes and autosomes since the prehistoric mutation?

By means of cloning, it will be possible to multiply intelligence (or other desired racial characteristics) after the image of the model cell. A development has got under way that is dangerous because it is attractive. Surgeons may point out that organ transplants could be carried out without problems, because there would be no immune reactions. One might also say a clan of clones represents the ultimate in inbreeding, but that is an erroneous view which presupposes that only one or a few types of men would be cloned. Once several types were cloned, they could mate with each other and 'normal' relations would continue.

We should be very naive if we imagined that the process would only be used in the positive sense. Apart from the extreme example of murderers and dictators being cloned, there is the possibility of production going wrong and resulting in monsters, because using the 'raw material' is so indescribably difficult. What would happen to the unsuccessful examples? They would be human beings too. Ethical and religious sensibilities command us to preserve human life. Every advance has its unconditional responsibilities.

Chance and danger are very close neighbours. On which side are the scales coming down? Should a strict ban be put on research into molecular biology and gene surgery? Apart from my view that we act under a compulsion to acquire knowledge, a research ban would have to be observed in every country and corner of the earth. Only comparatively small rooms are needed for genetic research, not great halls with thousands of machines and apparatus. Who is going to control them? Who will know if the ban has been universally beyed? Moreover, research has *never yet* been restrained from reaching goals that were ripe for discovery.

In addition to biological and ethical problems, there would certainly be legal problems to solve as well. Who is the testator in a series of clones? Who are the heirs? Where are the limits of direct descent. When everyone comes from one cell?

A biological time bomb is ticking away.

Nevertheless, I plead for the continuation of clone research in order to perfect the process and I want the rules for manipulations to be strictly guarded in special vaults. Faultless male and female cell nuclei, together with the surrogates necessary for bearing a child, should be stored in suitable substances and at life-maintaining temperatures. In case of catastrophe. This might be a cosmic catastrophe unleashed by a meteor passing too close to earth and emitting poisonous gases into the atmosphere. It might also be an atomic strike destroying a large part of the earth and causing radioactive emissions which would gradually damage human hereditary material. Then the human race could be fostered again by cloning—just like on the first day.

But should mankind be faced by such catastrophes for the first time, cloning could not be discovered and tried out *ad hoc*. Therefore the process should be immediately usable and tested in advance.

Cloning would not simply produce uniform types in a single large brood. Even though created after a pattern, there would be individuals, like their fellowmen produced *in vivo* in the tried and true tradition. They would resemble each other externally, they would also have the layout of the cell nucleus, but they would think and act independently, and like us they would be formed by upbringing and environment. The cloned men would receive new hereditary information and hand it on to a new generation. They would mutate and after a dozen generations the clones would no longer look as alike as peas.

Cloning is vital to our very existence in case of catastrophe, but it is also essential for the conquest of space. So I think the physiologist Lord Rothschild is right when he advocates the establishment of an international 'Commission for Genetic Control', so that research and practice keep in step. It is most devoutly to be wished that such a commission will be more effective than international organisations have been up to the present.

The German word for life (*Leben*) spelt backwards means fog or mist (*Nebel*). We should raise the mystical veil of mist with caution so that we can comprehend the reality of our existence.

Communiqué

The intelligent robot is on the way!

It will think independently and have an intelligence quotient far exceeding man's. It will be equipped with sensors which 'see' better than the human eye, for they will also see in the infrared and ultraviolet range. It will 'feel' more intensely, because its sensors function more sensitively than the human tactile sense. Its feelers—supersonic waves, radar, X rays—will 'feel' through walls.

The American scientist Marvin Minsky of the Massachusetts Institute of Technology, Boston, says:

> 'The machine will be able to tell a joke and win a boxing match. Once this stage is reached, the machine will develop at fantastic speed. In a few months it will reach the intelligence level of a genius and a few months later its power will be incalculable.'

Dr George Lawrence, scientific director at the Stanford Research Institute, California, has already linked human brains in direct contact with computers. The power of thought alone is sufficient to give the computers orders. The body which commissioned this Utopian-sounding series of experiments was the Pentagon!

The branch of research in the USA which has set itself the goal of creating an intelligent robot is called AI (Artificial Intelligence). The ultimate target is a robot which can carry out civil, military and scientific tasks in space and the depths of the sea quite independently.

Did intelligent robots exist in antiquity? The Sumerologist S. N. Kramer translated this passage from a cuneiform tablet:

> 'Those who accompanied the Goddess Inanna were beings who know no food, who know no water; they eat no scattered meal, they drink no sacrificial water . . .'

In the Sumerian Epic of Gilgamesh, Enkidu describes the guardian of the precinct of the gods:

> 'Not until I have slain this man, if he be a man, not until I have killed this god, if he be a god, will I direct my steps to the city . . . O Lord, who hast not seen this thing . . . thou art not stricken with horror, I, who have seen this thing, am stricken with horror. His teeth are like dragons' teeth, his face is like a lion's face . . .'

Sources: United Press International—S. N. Kramer, *History Begins at Sumer*, London, 1958—James Pritchard, *Ancient Near Eastern Texts*, Princeton, 1950.

3: Malta—a Paradise of Unsolved Puzzles

In the jet age the Maltese islands, 95 km south of Sicily, are virtually outside my front door.

I wanted to take a second look at something that every tourist stumbles over sooner or later, those strange 'ruts' in the stony ground with which all the Maltese islands are covered. In the 1975 *Lexikon der Archaeologie* (30), the entry under Malta reads as follows:

> 'More emigrants from Sicily came to the island around 3200 BC. An astonishing number of megalithic temples were constructed between 2800–1900 BC. The still extant temples, some thirty in number, exhibit a highly developed plan and superstructure . . . This population possibly followed warlike immigrants from western Greece . . . The strange "cart-ruts" belong to the same period.'

Even after detailed study of this odd phenomenon, no better word than rut has occurred to me.

Malta, the largest of the islands, with the capital Valletta, is some 25 km long by 12 km wide. The small islands of Gozo and Comino have their special attractions, yet Malta beats its lesser opponents not only by its size, but more especially by its unsolved puzzles: the ruts and the megalithic temples.

Sun, sea and weather have left their mark on the people and landscape of this group of Mediterranean islands. When you fly into Valletta you think you are landing in a sandstone cubist world. The squared buildings with their flat roofs lining the grid system streets lead into delightful pastel-coloured fields which might have been divided up by a ruler.

During the drive to the Malta Hilton in an ancient Ford, vintage 1954, the taxi-driver was lavish in his praise of the new Socialist government. 'We're going to throw out the

English and everyone else who's no use to us!' Whether I was interested or not, I was also told that Dr Dom Mintoff was a superman who would ensure that the people of Malta made constant progress.

I could not see many signs of this. Since my first visit eleven years ago, the holiday paradise with its fine hotels, beautiful streets, enticing shops and well-kept beaches had lost much of its glamour. In December 1974 the island became an independent republic and now it was advancing under the leadership of superman into the grey boredom of socialism. I could find little of the unique quality that travel guides and novels had once attributed to this 'paradise'. In a few days' time I knew that I would never come here for a holiday again. However, the Maltese fishermen still paint their boats all colours of the rainbow. For a moment I was reminded of Hongkong, although there are no junks in Malta.

Of course, the 'ruts' are just as well-known to the islanders as the Knights of Malta who turned the island into a European centre of culture at the end of the sixteenth

century. But the natives consider these 'cart-ruts' of little importance, as does the Maltese government, which does nothing to protect this unique feature. New building ignores the cart-ruts which are exposed defencelessly to wind and weather.

At some time or other every visitor will come across a rut or pair of ruts and as he steps over them, he may think fleetingly that they are unused stretches of an old railway line from which the rails were removed to melt down the valuable iron for other purposes. Perhaps the observer may even think that the markings in the ground *are* ruts made by carts. I do not know how many interpretations there are, I only know that none of them can be right.

The Maltese ruts are a unique prehistoric puzzle. Today there are still some hundreds of them on Malta and Gozo, but thousands of years ago the two islands were covered with them. When you look at these furrows impressed in the ground, most of them parallel as they should be, the natural reaction is to think of ruts. But closer examination of these mysterious traces shows that they cannot have been ruts in the normal sense of the word.

The tracks of the two parallel furrows are not only different from rut to rut, but also vary in the course of a single stretch. This is very obvious near Dingli, south-west of the old capital of Mdina, where the ruts mass together as if they were in a large shunting-yard.

They really are 'strange ruts'—even the archaeologists are astonished by them. They run through valleys, clamber over hills, frequently several side by side, then they surprisingly unite into a two-track stretch, only to take sudden and incalculable curves or run straight into the depths of the Mediterranean. Others again end abruptly at sheer cliff edges. In these places the rocks and the ruts must have crashed into the water together.

There is a wealth of different tracks. They are from 65 to 123 cm wide. The furrows are frequently over 70 cm deep.

The tracks of the 'miraculous' ruts run parallel — just like 'ruts' do!

Near Mensija one rut runs in a curve over the spur of a hill and cuts 72 cm deep into the limestone ground.

To deal with the cart-rut theory: if a cart ever covered this ground, it could not have taken a curve because of the great depth of the ruts. Either the axle-tree would have vanished in the deep imprint or the axle must have been at least 72 cm high, in other words the wheel must have had a diameter of nearly 1.5 m. But such a giant wheel could not have been coaxed round the curves; it would have got stuck or broken down. The independent suspension of our modern cars was unknown in those days, quite apart from the fact that the 1.5 metre-high wheels, which must have been like the wheels on an excavator, could not have manoeuvred in the relatively narrow ruts.

A sand-box game shows how ridiculous the idea of carts using the Maltese ruts is. Depth of rut . . . 72 cm, width at deepest point . . . 6cm. The curvature of the arc would correspond to a closed circle with a diameter of 84 m (!). Put *one* cartwheel, the axle of which must be over 72 cm high, in a rut and move it in arcs, without sand crumbling off the sides! It is impossible for the wheel to run under these conditions. This game would be quite inconceivable if the edges were made not of sand, but hard stone! As every single-axled cart has two wheels which must run absolutely parallel in the furrows, we can bury this theory for ever in some deep archaeological hole.

But if, for the sake of amusement, we postulate a *twin*-axled cart, the game becomes even more impossible. For inescapable technical reasons, the rear axle and the rear wheels would have to trace a *narrower* track, with a smaller radius, than the wheels of the front axle. That is why lorries take a wide swing on tight curves. As there is no narrower second track on the curves, we can rule twin-axled carts out of the prehistoric car pool.

Near San Pawl-Tat-Targa four pairs of ruts join up into *one* rut, although they have *different* 'gauges' when they meet. Hocus-pocus. Not far from there, one rut crosses another, but they are of different depths. Near Mensija, the 'railwaymen's' work was slovenly. The rut is fully hollowed

The 'ruts' often curve mysteriously.

out and up to 60 cm deep, with a width of only 11 cm at the deepest point and of 20 cm at the highest point.

On many sections of the coast, e.g. at St George's Bay and

The grooves cut as deep as 72 cm into the stone; their width varies from 65 to 123 cm.

south of Dingli, the ruts run straight into the blue waters of the Mediterranean. Until recently it was assumed that the ruts would end underwater a few metres from the coast and so have originated at a time when the sea level of the

Mediterranean was lower. Wrong! Divers corrected this misguided theory and told scholars about their latest findings, The ruts continue in the stone to great depths *below* sea level. Surprising, but true!

Archaeologists, too, think that such a large network of ruts must have served a purpose. They looked for one. When the remains of a Roman temple were found near Tas Silg in 1970 and excavated, the spades struck the walls of an older Greek temple at a lower level. The archaeologists thought that was all, but the ground held another surprise for them. Massive monoliths were found at the next level. When they were excavated, the semi-circular façade of a megalithic temple was revealed.

As there is a lot of talk nowadays about monoliths and megalithic buildings, I shall describe them briefly. Monolith is the name given to individual artifically dressed stone blocks, such as the Egyptian obelisks or the free-standing menhirs (Celtic 'long stones') near Carnac in France. Megalithic buildings (Greek 'graves made of large stones') were erected with large blocks or slabs, or sunk into the ground. They also include domed graves.

Monoliths have also been called in to explain the ruts. The ruins near Hagar Qim were made of monoliths 5 m high and 1.05 m thick. Here there is another stone slab with these gigantic measurements: length 7 m, width 3.12 m, thickness 64 cm. A real monster!

With hindsight the archaeologists explained that the ruts originated when the monoliths were transported to the building site—grooves cut into the ground by cartwheels!

Even a superficial knowledge of technology shows that this theory is a non-starter. For:

—The ruts have different 'gauges'. When there is a change of direction carts could not adapt to the new width.

—Gauges also alter in the course of a single stretch. Are the constructors of those days supposed to have had rubber axles at their disposal?

—Cross-sections of the ruts show that they do not cut into the ground at right angles, but narrow as they get deeper. If cartwheels made the grooves, the cross-section should be horizontal at the base. If anyone tells me that the wheels had pointed wedge-shaped extremities, I should simply retort that heavy loads such as monoliths could not have been carried. The grooves would have been cut deeper into the ground with every delivery. Finally, how large must the diameters of the wheels have been for the axles to clear the ground level. Such explanations are just weak evasions or stupid jokes. So how *were* the ruts made?

Let us try another version.

Did the builders of the megalithic layouts use animal-drawn sledges to transport their material over hill and dale? If the prehistoric inhabitants of Malta did, at some point in time, use this means of transport, which is ill-suited to the landscape, the ruts were not caused by these vehicles, for the same thing applies to sledge runners as to wheels, only more so. Sledge runners are fastened even more rigidly at the axle. They would have been useless in the bewildering confusion of rut widths and abrupt curves.

Another version:

To solve the difficult problem of transporting their material, the primitive inhabitants constructed a wooden 'fork', the two prongs of which scraped over the ground, while the narrow leading part was hung on an animal. They then fastened the monoliths on to the prongs. Excuse my mirth!

The 'fork' would have been rigid. The width between its prongs could not have been altered. Moreover we should still want to know what kind of wood could bear such burdens and what species of animal could stand the weights. The Maltese would have had to harness dinosaurs to 'forks' made of tempered steel. There was no steel, so it must have been wood that was as hard as steel. But neither the carrying capacity of the forks nor the cross-section of their wooden prongs can solve the puzzle of the narrow, pointed ruts.

Could forked poles have formed such deep grooves?

There is yet another fact to refute the idea of transport by carts, sledges or 'forks' drawn by animals. If animals had covered the same stretches for years and years, dragging a load behind them, they would have left their own tracks in

Are 'ball-bearings' the answer to the problem?

the ground. In other words, the paths of the animals which pulled the heavy loads would be as evident in the limestone as the ruts themselves. There are no hoof marks in between any of the Maltese ruts.

The Maltese used balls for their transport! In fact, hundreds of balls have been found in Malta. They are of soft limestone and come in different sizes. The biggest are 60 cm in diameter, the smallest 7 cm. Did the prehistoric Maltese invent vehicles which moved on ballbearings? Brilliant! Did they put the balls in the ruts and load the monoliths on to them? That would explain everything. Why the 'gauges' vary. Why the ruts take such bold curves. Why the ruts could intersect each other. Balls follow a track traced in the ground, no matter how wide it is.

Are balls the answer to the mystery?

Unfortunately not. All the Maltese islands consist of sandstone, limestone and clay, so the stone available is soft. And the balls are made of *limestone*. A weight of only one ton would squash them flat as a pancake or crumble them up like a snowball. Besides, balls regardless of size, cannot make pointed ruts; they always leave a curved channel. The ruts would be enlarged sideways, not downwards. But if balls burrowed 70 cm deep into the stone, they must have been gigantic, with a diameter of some 1.5 m, apart from the fact that they were loaded with an enormous weight. On top of that they would have had to overcome tremendous friction. What tremendous tractive power, what a monstrous thrust would have been needed! Discussion of this theory is pointless, as no balls with a diameter of more than 60 cm have been found so far.

No pictures of reliefs of carts or carriages have yet been found on Malta. But if the temple builders had used this kind of transport, they would surely have been portrayed, for there are other very ancient wall drawings and pictures on the island.

The ruts did not originate because of the temples, for if they had, the 'rails' would lead to the buildings and end there. But they don't! In a dense network they run past the temples in all directions, and are also found where no temples or other buildings rear their ruins above the earth.

The ruts have not been mapped or measured. It would indeed be a laborious undertaking. In many places they are overgrown and no longer visible on the surface, then they suddenly emerge again. Houses have been built over them and they have been affected by the detritus of the millennia that have passed over them.

No one knows what this maze of ruts was used for or who made it. Nowadays there is a lot talk about interdisciplinary research. Archaeologists ask physicists, chemists and metallurgists to collaborate in the solution of puzzles like this. There is no question of such cooperation on Malta

It is obvious that analyses of the ruts could fill in blank patches on the map of ignorance. Did stone balls, wooden 'forks' or cartwheels leave traces when they rubbed against the stone? No matter how the means of transport was built in days of yore, did it press minute organisms into the limestone or clay pores? Are there fossil pollen remains to be found that would help to date the origin of the ruts?

Today we have all the technical aids needed to examine the ruts running deep into the sea under the academic magnifying glass. Why doesn't someone do it? Can and should we leave such a fascinating puzzle from our primitive past unsolved? Normally we westerners are so terribly sensible and thirsty for knowledge. Why not in a case like this?

It would be surprising if the fairy story of the calendar were missing from our survey of the various theories. It crops up in connection with the Egyptian pyramids, the hanging stones of Stonehenge and the 'landing strips' on the plains of Nazca in Peru. The assumption that the Maltese ruts, too, were part of a larger-than-life calendar system is the silliest 'sensible' answer to a wide open question.

Everywhere that these archaeological tear-off calendars stand for sale, they consist of fantastically large buildings and complexes which are outside the range of vision of a man walking at ground level. Did stupid Stone-Age men lay out these gigantic calendars because they wanted to find out when spring began or autumn was coming? Nowhere is it recorded that prehistoric peoples practised agriculture on a large scale. If they *did do so*, with their small populations, they would not have had enough men or time, not to mention the backbreaking work and superhuman efforts required, to set up central calender stations of the kind we posthumously attribute to them.

Although my critics like to accuse me, even when they know better, of considering our early forefathers to be limited and incapable of personal achievement, I should like to take this opportunity of putting on record that I consider *all* species of *homo sapiens*, since he has existed on mother earth, to have been far too clever to have needed various kinds of so-called stone calendars to determine the change of the seasons. Our ancestors knew perfectly well from their observations of nature that spring followed winter, that the summer meant sun and the autumn cold.

In case I forget I should mention that scholars, as in other places, have speculated that the Maltese ruts were associated with a religious cult. We are not told what *sort* of cult it was. We do not learn for which gods, with their bird's eye view, the network of ruts was intended. If this cult theory is to carry any weight at all, I should dearly love to know what 'morse code' message these religious ruts were supposed to send and which airborne gods were supposed to be recipients.

I have already quoted from the Lexicon of Archaeology that the megalithic temples are supposed to have been built around 2800–1900 BC and that the origin of the ruts is also dated to this period, i.e. at the end of the Neolithic and the beginning of the Early Bronze Age.

That is all unsubstantial.

Excavations and cave investigations showed that Malta had long been colonised around 6000 BC. Statuettes of mother goddesses are 5000 years old. The Sicilians came in 300 BC the Phoenicians around 1400 BC.

I have not found anyone who says that the ruts should be attributed to the Neolithic Age, but scholars do claim that they originated in the Bronze Age. Yet even this comparatively 'recent' dating cannot be right. Were the people of the period intelligent fishes? Or did they knock up bronze diving suits, with snorkels, wooden airpumps and transparent eyepieces, so that they could work at making ruts on the seabed?

To be brutally frank we are taking flight into the unknown. Not so, say many archaeologists; the rut network existed more than ten thousand years ago, when the coastline now under water was part of the mainland! But listen to me—what tools were used to cut, mill, drill or bore the grooves, which run for kilometres?

The answer, of course, will be flint tools. That sounds reasonable, for flint, which is harder than limestone, was the material from which implements were fashioned in the Early Stone Age. But geologists have not found any flint in Malta or its neighbouring islands! Are we supposed to believe that flint was imported in the enormous quantities that would have been needed for the rut system in the Stone Age (!), which is not noted for international trade?

Others say, that's all wrong, Greek or Phoenician immigrants actually planned and built the network. Why not? But this idea does not get us out of trouble, either. According to all known records, immigrants who intended to put their knowledge to use in their new homeland always brought it with them and had already used it in practice. There is no trace of ruts in Sicily or Greece.

How grotesque the contradictions are! It is said that the megalithic temples were erected long before the *later* immigrants arrived. It's enough to drive one mad! If the temples were already built when the ruts were made, they cannot have been routes for the transport of building material. Moreover, the theory that the ruts date to around 5000 BC does not take into account the fact that with insignificant variations the Mediterranean has maintained its present level for at least 10,000 years. In other words, the last immigrants 'from western Greece' are out of the question as rut builders.

I look on the Maltese cart-ruts as an exemplary case of a wrong archaeological approach. There is a wealth of explanations but if you scratch the surface of the fine façade, the varnish comes off and the whole threadbare edifice is laid open. Nevertheless, any old theory gets into some specialised book or other where it is enshrined as the last word on the subject, so regardless of what book a reader lays hands on the opinion published in it is *the* solution. That is how schools are formed which stubbornly propagate *their* opinions, because they are unaware of or will not tolerate anything different. The main thing is that a question can be erased as answered. A genuine and final solution to the puzzle is less important.

It is clear that something happened in Malta in prehistoric times that never recurred anywhere in the world. The island must have been a centre for someone and something. Another theory is that metal alloys were poured into the deep grooves. But this view falls down because the grooves must have originated in times when metal was not yet worked. The post-Ice age alteration in the level of the Mediterranean proves this.

The following idea is worth a thought, too. Did saurians unknown to us mark the ground in prehistoric times, did they etch the strange tracks? The fact that the ruts run in parallel straight lines, with occasional curves, refutes this assumption, besides saurians do not leave regular shapes, but tread the earth at random.

We can also set aside the possibility that the ruts may have been open water conduits. No one can deny that water always runs downhill to the lowest point. Yet the ruts run over hills and valleys. Water can only be led uphill if the source of water pressure is at the highest point *and* the water is in a pipe. No pipes or remains of pipes have been found in Malta. If anyone in those days was clever enough to build a system of conduits, he would have chosen the shortest distance between two points and not built it in bizarre windings or zigzags.

A drainage system of this size would also have had to serve irrigation purposes, but the islands were poor and rocky at all times. Nothing flourished. Humus had to be imported! Only 40 years ago captains who watered their ships at Malta had to pay in humus rather than cash!

Are there any other explanations to examine?

Could it be that a natural product, unknown to us today, was cultivated in the ruts? Were silkworms farmed in their depths? Did some prehistoric algae culture exist that was used for food? We can erase those questions. To whom was

Could acids eat such regular 'ruts' into the stone?

the mass production from the extensive network delivered? The natives could not have used it all and as I have said nothing is known of a prehistoric merchant marine. Farmers as bright as that would have laid out their plantations more rationally—next to each other, not straggling over hill and dale.

Could the ruts, with their frequently bizarre twists and turns have been an obsolete form of writing? This attractive speculation falls down when it comes to the 'characters' lying under water. Who was supposed to read them *there*?

If the absurd 'writing' was actually 'written' in the limestone ground *before* the Mediterranean reached its present level, the 'readers' would have had to be able to fly! Otherwise the writing, which covers a large area, would have been illegible.

Let us put another Utopian-sounding idea under the magnifying glass. Could a metal alloy have been cast in the ruts to serve as a gigantic antenna? Who could have been the constructor over 10,000 years ago, when metals were unknown? Not even the megalithic temple builders.

Have I overlooked anything essential? Has the most important thing escaped me? I don't think so.

The megalithic temples are stone testimony that the earliest inhabitants of Malta worshipped their gods zealously and spared no effort to show their deep admiration of the heavenly figures. As you know, it is my view that the 'gods' were not fictitious figures, not the products of unbridled imagination. At some time they were real and physically very active beings.

I ask myself whether 'my' gods chose Malta as their goal in prehistoric times, whether they achieved something there that made the Maltese labour to cut stone signs in the ground in memory of or as homage to the extraterrestrials.

The General Manager of the Hilton Hotel in Malta, Mr de Piro, supports the original and intriguing idea that the ruts really were chiselled out of the rock by human hands. 'Why all that effort?' I asked him.

'You know that if you trace a track ahead of a domesticated animal, say an ass, a horse or an ox, it becomes a creature of habit and will cover the special stretch on its own, just as it finds its own way to its stall. Perhaps an apparatus that touched the track was hung round the animals' necks and then they went up and down the same stretch year in year out. That certainly would have left traces over the centuries.'

There is something attractive about the idea, but it did not convince me when I was on the spot.

The limestone ridge of San Pawl Tat-Targa lies between the towns of Ghargur and Naxxar. Exposed to wind and weather, heat and cold, the whole hillside is covered with ruts. A parallel track runs up over the ridge, makes an abrupt curve downwards and loses itself somewhere among the houses on the beach. No less than six more pairs of ruts cross the curve. But the points of intersection are not made in such a way that an animal could proceed further alone and independently. The crossings either end abruptly in a right angle or they are a different depth from the curved ruts, and sometimes they are as much as 81 cm deep, in which case any animal would have broken its bones. Lastly, some stretches

A drainage system? Can water run uphill without pipes? Where does water come from when there are no springs? And no water-towers?

simply peter out, after diminishing gradually. What happened to the animals then? And where are the paths made by their feet? The grooves themselves cannot have been made by animals, they are too deep and too pointed. When the ruts became level with the surface of the ground and vanished into thin air, so to speak, the animals must have ended up somewhere. Were they picked up by helicopter?

The theory that ruts may be connected with the building of the megalithic temples is most attractive and at first sight spiced with a touch of logic. The thirty megalithic temples which were built on the island of massive stone blocks and menhirs are edifices of Olympian dimensions. Malta has an area of only 247 sq km, Gozo an area of 76 sq km. New datings were made using the radio carbon method on the remains of wood found in the megalithic temple of Hagar Qim. The building was dated to 4000 BC. The ancient Romans, whose oldest early Ice-age settlement was founded in the first millennium BC, were not yet active at that time and even the ancient Greeks' first land conquests date to around 1200-900 BC. This supersedes the view that civilisation spread in the direction of Europe from Sumeria by way of Babylon and Egypt. The Maltese architectural miracles originated in the Stone Age!

Although I do not trust the C–14 method too far, because it is based on a constant relation to the radioactive C–14 isotopes in the atmosphere and finds of wood or bones say nothing about the time at which an archaeologically significant building was erected, I am still glad that Hagar Qim was dated to 4000 BC. That at least fixes a minimum date. We can conclude that the temple is not more recent, but may well be older insofar as the dated wood remains were not left by the builders of the edifice.

In addition Hagar Qim is still a Maltese dialect word which originally meant something like the 'prayed to stones'. Local archaeologists assume that the temple of Hagar Qim was dedicated to Phoenician divinities. Circa 4000 BC? Strange. There are no indications of the existence of the

Trails left by pack-animals?

people from the 'purple kingdom of antiquity' (31) at that early date.

If the routes were connected with the temples, the strange lines should logically lead to them, but that is just what they do not do. The thirty temples are scattered all over the island and the ruts run past them in the same random fashion.

There is the large complex of Tarxien near the town of Paola. Hagar Qim is only a few hundred metres from the temple near Mnajdra on the south coast. The temple of Skorba rises in the middle of the village, while the monumental prehistoric building of Malta, the temple of Ggantija, is to be found on the neighbouring island of Gozo. The great question is which came first—the megalithic temples or the ruts? It is as hard to answer as the old chestnut about the chicken and the egg.

There they lie, the gigantic monoliths. The millennia have passed over them. Have weathered and split them. When we look at the ruts, we think about how much rain has poured down on them, how many hundreds and thousands of times cold and burning heat have worked on them. Did they originally lie deeper in the ground? Have they been pushed upwards? Only one thing is definite: they were there before the Mediterranean reached its present-day level. Does this mean that the temples, too, should be dated to before the Ice Age? We do not know, but the assumption seems likely. But

The monolithic temple of Hagar Qim – but no ruts lead to it or to any of the other massive stone sanctuaries.

before I put forward my daring theories on the subject, I want to talk about another unique feature that Malta has to offer, besides ruts and temples.

South-east of Valletta in Saflieni, close to the town of Paola (12,000 inhabitants), the Hal Saflieni Hypogeum astounds the visitor. Hypogeum comes from the Greek and means 'underground room' (hypo = under, gaia = earth). In archaeological literature, the concept hypogeum stands for subterranean vaulted burial chambers and religious sanctuaries.

The house through which you reach the underground chambers differs from the others in the street by a massive door of four rectangular columns topped by heavy stone beams. A marble tablet on the wall reads: HAL-SAFLIENI PREHISTORIC HYPOGEUM.

The literature about this curiosity had been extravagant in its praise. When, after a long walk through streets that shimmered in the heat, I stood before the monumental limestone portal, I debated whether I should go in with my two heavy camera cases which were cutting into my shoulders like lead weights. For days a pitiless sun had been beating down on the dusty arid island. It was an atmosphere that considerably diminished even my boundless appetite for research. My shirt and trousers were stuck to my body. Finally I decided to enter. A quarter of an hour in the cool of the hypogeum would do me good. I stayed there all day and soon forgot that I had not been in a very good mood.

The entrance is at ground level; then you go three storeys down into the earth. A stately two-metre-tall Maltese emerged from the gloom of the hall and with weary inevitability took my camera cases from me. He could see that I was irritated and ready to argue, and quickly said: 'No cameras!', and in case I did not understand English added: 'Défendu!'. He stowed my cameras in a wooden cupboard in front of which he cut an imposing figure. 'Voilà!'

You enter the Hal-Saflieni Prehistoric Hypogeum through a doorway with a massive stone lintel.

Even today I don't really understand why you are not allowed to take photographs in many musuems. They could ask for a fee, obviously, but that cannot be the reason, for I was ready to pay anything they wanted in the Musée de l'homme in Paris, but without success.

I often suspect that the guild of archaeologists dislikes objects being illuminated from viewpoints which do not have their blessing.

Having learnt from experience that a lavish tip sometimes works wonders, I pressed two Maltese pounds into the giant's palm. He took them all right, but he would not even part with one camera.

He lowered his head to my 1.68 metres and whispered importantly: 'Sir, this is a holy place!' Of course, if this was a holy place I would have to be very obedient, otherwise there would be nothing but trouble. And in the back of my brain lurked the thought that if there was anything worth photographing, I would manage it somehow.

The camera custodian clapped his hands and a really impressive figure, a few centimetres taller than his colleague, emerged from a small office. I had fallen among giants. The second one was a good deal younger than the first. He wore a red silk scarf round his neck and a Basque beret on his head. As eagerly as any guide worth his salt, he addressed me in a polyglot lingua franca from which I gathered that English was his best language and that he had two attractions to offer—the so-called Museum and the Hypogeum which had brought me there. Museum is an exaggeration as it consists of four glass cases built into the wall. I wanted to see them both. When I had given the young giant another two pounds and asked him to explain things in English, he led me to the small showcases.

The treasures which were found by chance in 1902 during the construction of the house in which we stood were neatly displayed. Had it not been for these rather dreary artefacts, it is unlikely that the Hypogeum would have been discovered. I call that a lucky coincidence.

I was just able to make out *the* object about which I had read and which especially interested me before my giant dragged me to the staircase. It was the 'mother goddess', a terracotta figurine some 10 cm long, which is also called 'Sleeping Woman' in some books. She reclines in a shell which rests on four feet; her thickset body is wrapped in a garment which I can best liken to the shell of a tortoise; she supports her heavy head on her bent arm; her legs are short and stocky.

I prick up my ears when it comes to sleeping mother goddesses, especially when they come from the Neolithic Age. Why were Stone-Age artists so keen on depicting mother goddesses? What does mother goddesses really mean? Are the figures meant to be mothers of the gods? That is sheer nonsense. In the world of ideas of the Stone-Age artist the gods had no dependants, no family, no mother.

Stone-Age mother goddesses like the duplicate I was

looking at (the original is in the National Musuem in Valletta) have been found at La Gravette, Laussel and Lespugue, France, at Cukurca, Turkey, at Kostyenko, the Ukraine, at Willendorf, Austria and Petersfels, Germany.

Naturally the name 'mother goddesses' stems from our time. Who knows whether the Stone-Age artists ever conceived of the figures as 'goddesses'. Our clever attributions may make cataloguing easier, but I venture to doubt if they always get the *meaning* of what is represented. Never mind! These sculptures with their emphatically female and clearly pregnant attributes must have had a specific significance. Otherwise such figurines from the same period would have been found in so many places in the world. We shall see . . .

While my giant was guiding me down the stone staircase, he told me that what I was going to see had been found by chance at the turn of the century. I knew that from the literature. It was news to me that the original entrance to the underground world—a stone slab with a square hole in

The 'Sleeping Woman'.

it—used to lie on a hill above the harbour. When a wall was built there, the entrance was walled up.

My cicerone, who had a great gift of the gab, climbed gingerly down the spiral staircase, although he had done it thousands of times before. The further down we went, the quieter he became. In the end he only whispered when I asked him a question.

When we reached the main hall of the middle storey, I exclaimed: 'This is fantastic!' and asked; 'Why am I the only one here?' 'The Maltese don't come because they're afraid of the oracle. The hotel porters do send us tourists, but it's out of season now,' he whispered in my ear.

If the dating is right, we are told that over 6,500 years ago believers came to this hall to have their dreams interpreted by the priests in the next-door oracle room. I had read about the powerful acoustics, but I could hardly believe that softly spoken words would get louder, gathering volume and echoing through the hall. As if he knew what I was thinking, the young giant took me by the hand and led me to the niche. There he uttered long-drawn-out sounds into an ellipse hollowed out of the stone:

'Ooooohhaaaa' and 'Uuuuuhhiii!'

The giant's cries rumbled through the hall and bounced back off the walls as if amplified by the hi-fi system of a noisy discotheque. Even when he whispered, there were soft echoes from every niche and corner.

I simply had to try it myself. I stuck my head into the ellipsoid 'mussel' and spoke a long-drawn-out 'yes' into it. The higher I raised my voice, the more bizarre the resonance. If I lowered it to a sonorous baritone, it vibrated and echoed back from every corner. It did not escape me that the effect was particularly clear at a certain part of the ellipse. I directed my voice at it and reached the conclusion that concealed in the rock behind the 'speaking mussel' is a hollow space that acts as an amplifier, like the resonance box of a guitar. I assume, for of course you cannot see them, that

A flight of stone steps leads down 11.5 m under the earth.

hollow spaces branch out in the rock, transmitting the sounds and letting them out again in other parts of the hall.

As there were no ladies present, I was unable to carry out a test. Apparently the miraculous amplifier only works when a male voice speaks. Even if a woman raises her voice, the acoustics fail to oblige. Obviously I shall have to visit Malta again with a lady to whisper in my ear.

On my travels I have seen ancient sites that made a tremendous impression on me: pyramids and royal tombs in Upper Egypt, megalithic monster works in Turkey, the fortress of Sacsayhuaman above Cuzco, the 'water conduits' of Tiahuanaco and the gigantic statues on Easter Island, to name only a few examples. But the Hypogeum took my breath away. It was different from all the others.

Corridors, chambers and paths branch off from the great hall, niches and small chambers, two of which have painted ceilings, join on according to a well-thought-out plan. The niches and columns on which the dome of the hall rests are

This 'mussel', an ellipse hollowed out of the rock, acts almost like a microphone.

worked in faultless megalithic building style with clear-cut lines and sharp edges on the massive stone blocks. Even the dome is composed of curved monoliths.

'Did Stone-Age men do all this?' I asked my giant, who was enjoying my astonishment. He took off his Basque beret, twisted it in his hands and answered after a lengthy pause:

'They say that all this was cut out with hammers . . . '

By 'they' he meant archaeologists. One could sense from his answers that he had his doubts. Seeing the caves daily, he must have formed his own ideas about whether his early ancestors could have carried out this Titans' work with hammers.

As I was allowed to use my bright flashlamp, I could easily see that carving out columns, niches and sections of the dome was a masterly achievement. The monoliths that form

Columns, niches, curved dome sections – made of monoliths!

the niches rise from the stone floor without joints and they are of the same stone as the floor. Like crossbeams in a precisely calculated construction, more monoliths lie on them and they in turn are topped with monoliths curved into the shape of a dome.

What kind of oracles were uttered down here? Three, four

or five thousand years ago? The Phoenicians and Greeks did not consult the oracle. The sanctuary was covered over for thousands of years and hidden from the eyes of the immigrants. Graves that were found here are dated 1000 years earlier, to around 2500 BC, and the Phoenician and Greek invaders can safely be accepted as arriving between 1400 and 800 BC.

My lanky guide led me to a niche three steps lower down in which images of the gods may once have stood. He pointed to a hole in the ground that was closed with a stone slab. I learnt that there were a number of such openings and that excavations in the holes had revealed human and animal skeletons. No one knew whether the men and animals had been sacrificed. Even some thousands of years after the event the idea is horrible enough, but it is to get even more gruesome.

The middle storey in which we were is about eleven metres below ground level. We went down another seven steps. Now, at twelve metres, we were at the deepest point of the three-storeyed prehistoric complex. One last step and we stood in front of a rectangular dungeon. According to legend, it was used to dispose of unwanted intruders, to dump murdered enemies and for human sacrifice. Men who had volunteered to die went down there and graverobbers fell into fatal traps. The dead—7000 skeletons were found down there—guard their mysterious secret.

I read in a guidebook:

> 'The underground temple and oracle of the *unknown primitive population* consists of several passages and rooms and is excavated or cut out of the rock three storeys deep under the earth.'

To this laconic statement we should add that enormous quantities of flint which did not and do not exist on the island must have been used for the hammers—as for the ruts.

The Stone Age gets its name because men worked with stone tools. Metal was unknown then. But neither was there any flint, which is harder than limestone, on Malta. Nothing, absolutely nothing is known about fleets or ships which could have brought flint to the island from overseas. They did not exist.

If we stubbornly consider the question of material as solvable, the main puzzle remains. For what reason was the Hypogeum built three storeys deep under the earth? There is also the problem of the highly skilled architecture. The goal must have been fixed from the first hammer-blow on the stone, the continuation of the work planned and the intervention of the stonemasons coordinated.

Let us imagine the work of a Stone-Age architect, just for fun. He scratched a few hundred sketches on palm leaves, following a model inspired by the gods in a dream. How else would he have thought of the daring construction of an underground dome for which there was no precedent?

Our bold Stone-Age architect planned his layout three storeys deep under the earth. Where did he get the necessary knowledge of stresses from?

What scaffolding did he give the stonemasons for the straight and curved monoliths? They had to bear their own weight and that of the storey above!

When our audacious architect laid his astonishing plans before the masterbuilder, the frustrating question of the necessary tools hung in the air. Given the existing state of Stone-Age implements, there was no answer. What a pity!

The building was greatly improved by the acoustics which I have already mentioned *and* by a first-class air-conditioning system! The Hypogeum has a built-in one. Whether a single person like myself visits the three-storeyed underground house or hundreds of tourists walk through the halls, there is little change in the temperature. Yet everyone knows how quickly the air in closed rooms warms up when people give off heat like living radiators. The system in the

Hypogeum at Saflieni is as sophisticated as that in the underground towns of Derinkuyu in Turkey where the temperature is constant in summer and winter in all 13 (!) storeys below ground level.

In the case of Derinkuyu, scholars have agreed for the sake of simplicity that the sophisticated towns were built in post-Christian times (as if heating engineers were two a penny after Christ!). That is not true, but the dating must do as an explanation for the first-class ventilation system. We cannot take this easy way out in the case of the Hypogeum; its Stone Age origin is undisputed.

If the construction and stonework are puzzling and the acoustics a phenomenon, the Stone-Age air-conditioning fulfils requirements one can only describe as astounding.

It is thought that the hypogeum was built in three stages. Scholars think so, because halls and niches differ from one another architecturally. On the upper level, natural hollows in the rock were simply enlarged and smoothed, whereas in the main hall with its subsidiary rooms in the central storey a hitherto unexplained (artificial) megalithic method of building undoubtedly stamps the layout.

This explanation has a weak spot. The different techniques must have been used *simultaneously*, because both acoustics and ventilation system embrace *the whole Hypogeum*. Therefore the first architect and his successors must have had a clear idea of the finished complex from the beginning. Subsequent corrections or 'installations' cannot be made when a building is fashioned out of the stone.

To me, ruts, temples and Hypogeum are proof that 'gods' took a hand here.

I have to make a statement which will be superfluous to anyone who understands my theory, because I want to anticipate a remark that my critics will make as sure as eggs are eggs. I do not claim that 'gods' were at work here, laid out the ruts, erected the megalithic temples or built the Hypogeum. But I do speculate that 'gods' or their descendants were familiar with tools and dominated techniques which the

Three storeys under the ground: a vault cut out of the rock (note the upper part of the photo), with curved monoliths.

Stone-Age men made use of. Obviously it is also possible that the early islanders worked zealously at the gods' request to make the ruts, without knowing *why* they were doing it.

Is there a connection between all these apparent contradictions? Can 'gods', men, ruts and temples all be brought under one hat?

Homer described the adventures and misadventures of Odysseus, King of Ithaca, over a period of ten years. Driven on to Cape Malea at the south-east tip of the Peloponnese by a fierce gale, he and his ships visited the island of the Cyclops, the one-eyed giants. They were the builders of megalithic walls which are still described as Cyclopean masonry today.

It is often surmised in scholarly literature that the island of the Cyclops was present-day Sicily. It may be so, but not necessarily.

Malta and its four small satellite islands are only 95 kilometres from Sicily. Anyone who studies the megalithic buildings carefully will share my impression that giants did the work. Were they the 'inventors' of Cyclopean masonry?

One of the Cyclops, the giant Polyphemus, held Odysseus and twelve of his companions prisoner in a cave, the entrance to which he blocked with an enormous stone. Polyphemus could leave the cave at will because he could pull the stone away, but it was too heavy for Odysseus and his men. Polyphemus was the one-eyed son of the god Poseidon. And all the other gods on the island were also sons of gods!

Is there a mythological reference to what was once reality? Did giants live on Malta in the very remote past?

No one can deny that giants *did exist* at one time. Early traditions tell us vivid stories about them and ancient texts stubbornly assure us that they were descendants of the gods, 'sons of heaven'.

The ruined monolithic temple near Mnajdra – built by Cyclops?

In the 14th chapter of Enoch, who according to Genesis (5, 18 *et. seq.*) was in immediate touch with God, we read:

'Why have you done like the children of earth and begot giant sons?'

In Genesis (6, 4), it says:

'. . . the sons of God saw the daughters of men that they were fair; and they took them wives . . . ' There were giants in the earth in those days . . . mighty men which were of old, men of renown.'

Chapter 100 of *Kebra Nagast*, the Ethiopian work, contains this paragraph:

'But every daughter of Cain with whom the angels had consorted became pregnant, but could not give birth and died. And of the fruits of their wombs some died, and others came forth; they split their mother's womb and came out at the navel. When they were older and grew up, they became giants . . .'

Lastly a line from *The Book of the Eskimos*:

'In those days there were giants on the earth . . . '

In Baruch actual figures are given:

'The All Highest brought the Flood upon the earth and destroyed all flesh and also the 4,090,000 giants.'

In my book *According to the Evidence* I included photographs of recent fossil finds of giant footsteps, the latest proof from reputable sources of the former existence of giants. I have no desire to repeat myself, but must at least mention in passing the documented existence of prehistoric giants, otherwise it will be: But Mr von Däniken, there never were any giants! People like to overlook what is not supposed to be true. *That is why* I have to mention it.

Let us spell out the little word IF, so pregnant with significance!

—If Homer was not merely giving rein to his poetic fancy, but actually handing on the core of real events in his Odyssey . . .

—If Malta was the island of the Cyclops . . .

—If the Odysseus landed there . . .
—If the Cyclops were descendants of 'fallen angels' and therefore of the extraterrestrials . . .
. . . then ruts, megalithic temples and Hypogeum had some direct connection with the gods *or* their descendants.

Why?

Let us remember that some ruts lead into the depths of the Mediterranean and so came into being *before* the last Ice Age, when the sea level was lower than it has been for millennia. Now from the point of view of classical archaeology, there were no *technologically* skilled peoples *at that time*. If, as a logical conclusion, it could not have been the Stone-Age inhabitants who bequeathed us the monuments we admire today, who was it?

Did gods or their descendants leave a sign of their presence on Malta? Quite apart from technical relics, did they set up semen banks in some as yet undiscovered spots, the entrances

The temple of Tarxien – Malta, isle of the Cyclops?

to which will remain undiscovered until they are opened by a lucky *coincidence*, as in the case of the Hypogeum? May the mother goddesses by the key to the last puzzle? Are well-preserved cells from the bodies of the former rulers of our planet waiting to be found under rocks or in megalithic sanctuaries? Will sarcophagi with mummified giants be excavated some day?

I do not want to be told that the ideas I am putting forward are extravagant, because they have a sound basis. Pharaohs and Chinese emperors, Incas and Japanese emperors, were versed in the art of mummification from the earliest times. So why, I speculate, should not giants, 'sons of the gods' and the first progeny of the extraterrestrials, have practised this art as well?

If the first *intelligent* men were scions of spacetravelling gods, they certainly acquired sufficient scientific knowledge from their heavenly fathers and perhaps even the order: 'Guard and preserve body cells. One day you shall use them to produce beings in your own image!'

When we were filing the photographs of our journey to Malta, my collaborator Willi Dünnenburger drew my attention to a characteristic of the Maltese 'mother goddesses'. All the statuettes are of *pregnant* women. Not only are the bodies swollen as if they were going to give birth to triplets, but the figurines have no thighs. The lower part of the body is unwieldy and fat. The calves are no longer recognisable; the swelling starts from the feet.

We could get round this observation by saying that the prehistoric sculptérs could not recreate the subtleties of the body, because they were too primitive. But that does not work, for shoulders and arms are very delicate and plastic in their modelling. Many figurines exhibit a hand with four fingers and outspread thumb held over the heart as if the woman wanted to express her pain or anxiety about giving birth.

Surely the sculptures hint that these wombs carried some-

Pregnant mother goddesses.

thing more than a normal embryo? Were the bodies of the pregnant women dragged downwards by the abnormal weight of the foetus? Did tissue, amniotic fluid and unnatural pillows of fat blot out thighs and knees? Could these poor creatures only waddle along a few weeks before giving birth?

Seen from this point of view, even the clumsy 'mother goddesses' have their value as proof of the existence of giants in the past. *Kebra Nagast* tells us about wombs split at birth because the foetuses had grown too big. A Sumerian cuneiform inscription from Nippur says that Enlil, god of the air, violated the child of earth Ninlil. Ninlil beseeched the profligate:

> '. . . my vagina is too *small*, it does not understand intercourse. My lips are too *small*, they do not understand how to kiss . . .'

I do not venture to speculate whether Enlil was an extraterrestrial or a first generation descendant, but it does emerge clearly from the Sumerian text that his body and its parts were too big for the normal-sized earth maiden Ninlil.

In the West there is yet another secret over which the veil of the millennia lies. Even the archaeologists admit they have nothing valid to say about it and that is a lot for a guild which professes to know nearly everything. I am talking about Brittany on the French Atlantic coast, 2,300 km from Malta, as the crow flies.

Gourmets are not the only people who go there to sample the celebrated fish and vegetable dishes. Brittany has been drawing travellers, today we should call them tourists, for hundreds of years because of the many thousands of menhirs.

When I was spending a few days in Brittany last autumn, I went for a walk through the menhirs on the night of the full moon. I felt as if I was on another star or in the primaeval landscape of our earth.

The menhirs or 'long stones' (the translation from Celtic) threw long ghostlike shadows. The colossi became a wild phantasmagoria. In the shadows I saw the pictures that did not exist. Sometimes men's faces, sometimes a mother with her child in her arms. Then lions, panthers, great crabs and spiders. They all slithered by me in the moonlight in oppressive silence. Prehistoric monsters, fabulous creatures, crouched in the distance ready to attack, yet when I came

closer they became again the giant moonlit stone relics from prehistoric times. I had been on a time journey into the past.

The long stones stand in an as yet unexplained arrangement, that is to say they are not erratic blocks, not survivals of an Ice Age. These columns in rows of three and twelve look like a petrified army standing to attention. The smallest of the stone 'soldiers' are at least one metre high. The giant among them, the menhir of Kerloas near Plouarzel, is 12 m high and weighs 150 tons. The biggest 'long stone', the menhir of Locmariaquer, lies broken on the ground. When whole it was 20 m long and weighed over 350 tons!

Near Kermario, 1,029 menhirs stand in 10 rows on a site about 100 m wide by 1.2 km long. Near Ménec there are 1,169 long stones arranged in eleven columns, 70 of which diverge and form a semicircle, a formation which is repeated at Kerlescan with different figures. Out of 594 menhirs, 555 of them in rows of 13, 39 form a semicircle. Near Kerzehro there are 1,129 stones in rows of 10, near Lagatjat 140 in columns of three.

The arrangement of the stones in Brittany has not yet been explained.

These data are not complete, but give an idea of the tremendous work that was carried out at some time in the past. The menhirs in Brittany have one thing in common with the megalithic complexes on Malta. Both of them must have been built *before* the last Ice Age, because, just as the ruts in Malta descend into the Mediterranean, whole columns in Brittany march into the depths of the Atlantic Ocean!

Often the natives have something revealing to say about the phenomena among which thay live.

Breton farmers, when I asked what these stones columns meant shrugged their shoulders and admitted: '*Personne ne sait*!' This admission of ignorance seems more rational than the Christian legend which others trot out at the drop of a hat. St Cornelius, who lived around the middle of the third century AD, was pursued by Roman legionaries. He implored Christ for help and with his assistance turned the Roman soldiers into stones, the biggest of them being the officers. So military ranks were preserved even in the menhirs. Fabulous.

Another implausible explanation is that the whole region of present-day Brittany was once the sacred country of the Druids. That may well be so, but the Druids, the priests of the Celtic peoples, had their great period in the century of Caesar, i.e. the last century BC. So if the Druids transferred their sanctuary inside the boundary of the menhirs, they could only have taken over an existing complex. Clever and thrifty of them!

We can also discard the claim that primitive nomads in the slumbering Europe of prehistory quarried the stones, then transported and erected them, in the same way as the eastern peoples who erected monuments in honour of their gods in Egypt and Babylon. Supporters of this version should understand (and know) that the megalithic age lay long, long before the epoch in which the Egyptian and Babylonian buildings originated. It dates back at least to the last Ice Age, to the time of the gods and the sons of gods.

What we marvel at in Brittany today can give only a vague idea of what was here ten thousand years and more ago. Men and nature, the great destroyers, have done their work.

In the middle of the last century the rumour that gold was hidden in the menhirs was rife in France. The gold seekers swarmed in, equipped with picks and sledgehammers. Gold fever shows no consideration. The long stones were savagely attacked. The remains of this great gold battle lie around and make a sorry sight. Menhirs that were once imposing are split in two, smaller ones have been smashed to bits. Today the government is trying hard to save the menhirs from further vandalism. The adults and children who clamber over the stones and do more damage every day take no notice of prohibitions, I find the carved initials with which stupid people want to perpetuate their memory particularly annoying.

As we walked through the ranks of the petrified legionaries on the fine autumn days, my daughter Cornelia asked me what I was asking myself: What was all this for? What do these thousands of stones, set up in rows of three, nine or eleven, mean? Were they tombstones ? No. However hard scholars have looked, no graves have been found at the foot of or under the menhirs. They have been found in dolmens, the megalithic tombs lying underneath earth mounds. There are more than 3,500 of them in France.

Did the menhirs once have roofs? Was Brittany dotted with gigantic halls? The different lengths of the stones refute this theory, as do the results of recent research which found no tenons or mortises for construction purposes. Besides, the menhirs are either too close or too far apart to make roofing possible. Where the stones are almost next to each other, it would hardly have been possible to move for menhirs. In places where the menhirs were far apart, there were no wooden beams or dressed stones long enough to connect them. Lastly, as the menhirs have survived the millennia—even though damaged—it should have been possible to find at least the remains of roofs. Nothing of the sort has come to light.

I have an acquaintance I meet once in a blue moon. He loves telling jokes, but has a very small repertoire. After the usual greetings comes his stereotyped question: 'Have you heard this one?' I quickly answer 'Yes', for he has certainly told it at least twenty times. I feel exactly the same when I read that the Breton menhirs, too, were part of a calendar. I can laugh at this theory as I do at my friend's jokes, without taking any notice of either.

This theory involves Celtic priests or their colleagues from megalithic times getting their sheeplike congregation to fetch thousands of stones and erect them in precise arrangements simply to find out which season was due from the 'geometry' of the stones or their shadows. The British astronomer Fred Hoyle thinks that the priests wanted to impress or frighten the people with the complexes. What, after the people themselves had dragged the massive stones into place?! The priests could certainly have made an impression by predicting an eclipse of the sun or moon, but here again the stone lines give not the faintest hint of the primitive observatory which would have been necessary.

My vehement objection to the calendar theory: simple predictions can be obtained with much less trouble. If such a complex made it possible to announce the advent of a spring tide, for example, it would be ridiculous. Spring tides occur twice a month owing to the attraction of the moon's mass. The seasons change according to an eternal rhythm. I venture to call my ancestors idiots if they assembled these stone masses to form a calendar simply for banal announcements like that. Basta.

To put it scientifically, an axiom is a self-evident truth or universally received principle. Theoretical assumptions can be deduced from axioms. In this way systems which are perfectly logical in themselves can be built up. I am going to

Could men of our build have transported stones like these?

allow myself to build up a theory from axioms:

First assumption: The menhirs in Brittany were not assembled by men of present-day stature. *Justification*: Weight and number of stones.

Second assumption: The menhirs were assembled *before* the end of the last Ice Age. *Justification*: Stone columns vanish into the depths of the Gulf of Morbihan.

Third assumption: The complexes were intelligently planned and built. *Justification*: The arrangement of the menhir is not a chance one.

These three axioms throw up new questions and permit conclusions. At the end of the Ice Age, who had the bodily strength and also the bird's eye view needed to set up such gigantic complexes of thousands of menhirs?

Giants!

Giants from early non-historical times are documented in traditions. Datings place them at the end of the last Ice Age and they would also have had the mental faculties and the strength needed to erect the menhirs.

To what race did the giants belong and what was their origin? Mythologies and religious traditions claim that they were descendants of the gods.

Another question; were the giants intelligent or stupid? The products they left behind will show if they were intelligent. Of course, we have to decide whether megalithic complexes like those in Brittany served an intelligent purpose or were merely some stupid occupational therapy.

The deliberate arrangement of the complexes alone proves that they were laid out to a *fixed plan*. Anyone who plans is intelligent. Axiomatic conclusions: intelligent giants extracted thousands of menhirs from the rock and transported the heavy stones to chosen sites where they erected them and arranged them in columns.

What goal were they trying to achieve?

The German engineer Rudolf Kutzer from Kulmbach has made a bold speculation. He thinks that the menhirs were arranged to form a signal antenna which could possibly have been connected to an amplifier for cosmic energy.

Are there any indications to justify this audacious claim?

Menhirs consist of quartz-bearing stone, occasionally with traces of iron. Quartz is one of the hardest minerals, consisting of the chemical element silica, Si 02.

Anyone who was ignorant of the special qualities of quartz will have learnt about them from the new generation of watches. In 1880, during their investigations into the electrical behaviour of crystals, Jacques and Pierre Curie discovered what is known as piezoelectricity which occurs when quartz crystals are subjected to pressure, pulling or turning in a specific direction. Using these minimal energies, watches can be kept going for a year or more.

Even as children we used small quartz crystals when we concocted simple radio receivers out of old boxes. We ran a very fine needle over the quartz, When a certain point was found, there was a noise in the earphones and we heard a nearby transmitter as if from a great distance. What happened to us little do-it-yourself boys?

Quartz picks up oscillations like an aerial and repeats them in concentrated form from a specific point. After a careful search, we had found *the* point via which the frequency of the transmitting station reached us—*without any electrical amplification*!

This special quality of quartz made Kutzer ask: were the menhirs 'charged' in some way. Were they 'stimulated' by some kind of energy unknown to us? Did they emit oscillations when connected with each other? Or did they receive oscillations from the cosmos? Questions unanswered as yet, but what do *we* know today about the possibilities of a *future* technology which may have been a *past* one for the extraterrestrials? As science is always striving to understand the past with present-day logic, everything that does not fit into the frequently ludicrous picture of inherited axioms goes by the board.

Strangely enough telephone wires throughout the world are still mainly carried on wooden poles, although wood is known not to be a durable material. It rots, decays and is highly inflammable. Nevertheless, wooden telephone poles are 'planted' in concrete sockets in many countries.

Archaeologists at work 5000 years hence:

Over hill and dale lumps of concrete with round holes in them protrude from the ground. Analysis clearly show traces of wood in the concrete pores.

The neatly arranged rows of concrete blocks lead to the conclusion that their ancestors (around the turn of the second millennium BC) practised a religion in which the rows of blocks had a special significance, otherwise the people of those days would not have taken such trouble to cover countries and continents with the heavy blocks. This explanation is opposed by another theory which says that the

Pillars supporting a former forest of antennae – like the ones in NASA's project CYCLOPS?

concrete rows were signposts, directional aids for large-scale migrations. Needless to say, the immortal calendar theory also rears its head again.

The only snag is that none of the theories can explain the clearly demonstrable traces of wood! So some scholars assert that torches were stuck into the concrete. Wood was dipped into inflammable fluid and ignited. Even before this theory can be included in the literature, critics object that it is absurd, because the concrete blocks are much too close together for communication by fire signals. When a young archaeologist says that they might have held telephone poles, there is a roar of protest. The men at the end of the second millennium were intelligent and possessed an astonishing technology. *Firstly*, radio was widely used, *secondly*, they would never have made telephone poles of wood, because other finds proved that they commonly used various metals. That is how, anno AD 7000 they will 'prove conclusively' that the concrete bases in the earth could not have been sockets for wooden telephone poles, because they did not exist around AD 2000.

Is our logic more conclusive?

As I write this, I can hear critics whisper mockingly in my ear: 'Didn't you say that giants of the megalithic age had assembled stones to form a giant antenna! If the giants, your giants, had any idea how an antenna worked, they would have used some kind of metal instead of the long stones!' How logical is this logic?

If *we* set up a forest of antennae *today*, as is planned in Project Cyclops, we should naturally use metal. The NASA Ames Research Centre's programme includes a vast site with 1500 directional antennae, each with a diameter of 100 metres. The giant antennae will rest on thousands of concrete sockets. But in thousands of years, even the metal of the Cyclops antennae will be rusted, reduced to atoms, washed by rain and blown away by winds. What will remain? The thousands of concrete sockets geometrically arranged on the ground. The ground, being very hard itself, will have protected them from corrosion.

Perhaps the technologists of generations to come will invent a system of transmission into and reception from space without metal antennae. Perhaps they will set a quartz mountain oscillating and use it as an antenna. Who knows? Did the first generation of the sons of the gods after the presence of the extraterrestrials who built the menhir complexes know such a process? Were they miles ahead in the use of the piezoelectricity in quartz?

Who knows?

My speculations are daring, the lines of communication between my axioms are not yet stable. If everyone practised intellectual modesty and paid tribute to Socrates the wise when he said: 'I know that I know nothing, and barely that . . . ' it would be ideal. As all previous speculations about the significance and purpose of the menhirs are out of date and illogical, new stimulating ideas pointing to both past and future can do no harm.

An apparently unimportant detail often suggests associations.

Nearly all the menhirs taper downwards from the top. We might think that the builders sharpened the long stones before ramming them into the earth. That is a reasonable idea, but I find it illogical on two counts.

The heavy menhirs would stand more firmly in the ground if they were rectangular all the way down than if they tapered from above. A construction expert confirmed this. Level ground plus a level stone base plus heavy dead weight guarantee stability. That is how we build high rise buildings today, using precast concrete supports. Pyramids taper towards their summit and have their greatest area at the lowest point of the platform resting on the earth. If it were otherwise, they would tilt to one side. The 'normal' menhir is like the pyramids. It has its largest cross-section below, it stands where it is inserted into the level ground. If it tapers from above, the base area is smaller and consequently its stability is lessened. Not only do the Breton menhirs exhibit tapering, they also have carved snakelike grooves *below groundlevel*. These are explained away as ornamental embellishments. *Underneath* the earth, where no one could see them?

Did these ornamental grooves once contain metal connecting the menhirs with each other? According to engineer Kutzer's theory, such connections would have been necessary for a 'forest of antennae' to function. The electricity in the quartz-bearing menhirs would have had no effect until it was concentrated. They were certainly not connected at their upper pointed ends where there was no support, whereas the peculiar decorations strongly suggest that they may have been connected at the bottom. Today only the grooves are left. No trace of copper (or any other metal), no trace of supports. Does that mean that the antennae theory belongs to the rubbish heap?

Think of the lightning-conductor. The part in the earth corrodes more rapidly than the part leading to the roof, although the latter is exposed to the elements. Why is the metal section in the earth destroyed more quickly?

Two pieces of *different* metals which are connected with each other form, together with an acid solution, a galvanic element. In every galvanic (electric) element, ionic currents flow in such a way that the 'baser' metal is decomposed in the electrolytic electrochemical series. The bigger the difference between a 'noble' and a base metal that work on each other in an acid solution, the more radically will the baser metal be attacked.

Magnesium (chemical symbol: Mg), aluminium (Al), manganese (Mn), zinc (Zn), chromium (Cr), iron (Fe), nickel (Ni), tin (Sn), lead (Pb), copper (Cu) and silver (Ag) form a series of metals which will always destroy the baser metal, in an acid solution. If we put a minus sign for the baser series and a plus sign for the nobler, it would look like this:

— Mg, Al, Mn, Zn, Cr, Fe, Ni, Sn, Pb, Cu, Ag +

Although metals plus an acid solution create a galvanic field, an ionic current, and metals dissolve in it, the 'acid solution' is missing in the case of metals lying in the ground! However, rainwater is mildly acid.

Now a corrosive current also occurs when one electrode is fixed in concrete and the other in the earth. The iron in the concrete becomes the cathode, the metal in the earth the anode. In the long term, the anode will be destroyed, dissolved, by the ions . . . With modern measurements of corrosive currents we calculate in advance how many grammes of a metal will be lost in a given time (33).

The logical conclusion is that if the menhirs with their rich quartz content were once connected with each other by metal, the metal underground would have been completely dissolved over the millennia, because the megaliths worked as cathodes. Another point is that such ionic currents can run from one monolith to another not only in a straight line, but also in circles. A single strong cathode close to the menhir groups would have been quite enough to dissolve metals down the millennia.

The speculation that the menhir complexes had a technical purpose is not so far fetched in connection with the dolmens, either.

'Dolmen', translated from Celtic, means 'stone table': dol=table, men=stone).

There is a great variety of stone tables: sometimes two clumsy megaliths support a gigantic stone slab, sometimes several slabs lie on small megaliths, sometimes more than ten covering slabs form dolmen corridors, sometimes the stone tables are covered over with artificial mounds—burial chambers.

Like the menhirs which have never revealed their purpose and significance, the mystery of the dolmens is also an unsolved case. Graves and skeletons which do not date from megalithic times were found under many dolmens. During the Bronze Age, later inhabitants of Brittany must have chosen already existing dolmens as their last resting place. If you ask the local farmers, they say the dolmens were 'giants' tables'. This answer conjures up another paradox: the dolmen corridors—too low for giants, but suitable for dwarfs, who could never have handled the heavy slabs. On the other hand, the very large free-standing dolmens from

Rostudel to Cap de la Chévre certainly remind us of 'furniture' for giants. Perhaps they, too, were previously covered with earth which was washed away down the millennia. We do not know. But if the menhir columns served a technical purpose in megalithic times, the dolmens certainly originated in connection with that system. Perhaps 'something' was hidden under the dolmens or they protected the environment from the something.

One day long ago, for unexplained reasons (that is not my hypothesis!), the constructors and builders of the megaliths disappeared or died. They bequeathed an astonishing puzzle to their ancestors, who even today do not understand what went on thousands of years ago. Will they disclose the secret tomorrow?

Communiqué

James Oberg, American spacetravel expert, is convinced that we can count on permanent Soviet colonies in space within fifteen years. Families would live in these orbiting satellite cities and their life would not differ in essence from life on earth. Oberg thinks that the first immigrants who came to America from Europe had to show far more pioneer spirit than future space dwellers will need.

James Oberg is not some crazy visionary. He is the expert on Soviet space travel at the Institute of Astronautics and Aeronautics, and he prophesies that:

> 'Spaceships with men and women on board will orbit the earth for so long that they will look on themselves as permanent inhabitants and no longer have any intention of returning to earth.'

The world-famous Russian astronomer Yosef Shklovski even goes a step further — he says that artificial biospheres will be set up in space during the next 250 years and that up to 10 milliard people will be able to live in them. Shklovski is not a fanatic, either. He is Director of the Radioastronomical Department of the Sternberg Institute in Moscow and a corresponding member of the Academy of Sciences. This highly qualified scientist assumes that raw materials from the moon, asteroids and other planets will be used to build the space colonies. Shklovski says:

> 'The erection of artificial worlds in space is inevitable. Once man's breakthrough into space has begun, it will be as irreversible as the discovery, colonisation and exploitation of new countries during the age of great historical discoveries.'

Shklovski is convinced that mankind will colonise the whole planetary system and thus begin an inevitable drive into other spheres of the Milky Way:

> 'Only the colonisation of space offers a long-term solution to the problems of mankind, as it has been proved mathematically that a strategy of limited growth aiming at global equilibrium cannot prevent a world crisis.'

4: History repeats itself

When I tore the page for yesterday, 7th December, from my calendar, I read this thought for the day:

'We need Utopias. Without Utopias the world would not change.'

The sentence is by the American writer Thornton Wilder, who died three years ago yesterday.

The maxim for today was supplied by Johann Wolfgang Goethe:

'Everything clever has already been thought; we must try to think it again.'

If a calendar compiler asked me for a pithy saying, I should like to write in thick black letters:

'All history repeats itself.'

Whatever the day, it would always prove right.

. . .

Garuda is the prince of all birds in Indian mythology. He is a multi-purpose bird, so to speak, for he is depicted with an eagle's wings and beak, but the body of a man. He must have been of powerful stature, for he served as the god Vishnu's saddle-animal.

Extraordinary faculties were ascribed to this remarkable bird. He was highly intelligent; he acted independently, waging wars and winning battles on his own. Even the names of his parents are known. They were called Kasyata and Vinata. Mother Vinata laid the egg from which Garuda emerged. In other words, everything began quite normally. On the face of it.

His face was white, his body red and his wings were

Garuda, prince of all birds.

golden. He would have cut a good figure in an ornithological book . . . but he would not have fitted in!

For when Garuda raised his wings, the earth quaked. That was when he began his journeys into space.

In addition he had one phobia; he could not stand snakes. But he had good reason for this.

His mother Vinata was kept prisoner by snakes when she lost a bet. The snakes promised to release mother if her son brought them a bowl full of ambrosia, the food of the gods that conferred immortality. The brave son made every attempt to comply with this condition, although ambrosia was only obtainable on a divine mountain which lay in the middle of a fiery lake. But Garuda even had a bright idea for coping with this precarious situation. The myths tell us that he filled his red body so full of water from the neighbouring rivers that he finally managed to extinguish the wall of flame and reach the mountain of the gods. But the mountain was teeming with fire-breathing snakes which tried to stop him landing. Once again Garuda had a bright idea; he sent out whirling clouds of dust. The snakes could no longer see him. Then he hurled 'divine eggs', which tore the snakes into a thousand pieces. He is supposed to have slit the tongues of some snakes which came too close to him. We can understand that.

Adventurous as this strange bird was, immediately after the liberation of his mother Vinata he set off for the moon! But it was in the possession of alien gods who did not want him on the moon at any price and so gave battle. However Garuda's body was immune to the gods' weapons; they could not harm him. Garuda was invulerable. When the moon gods realised this, they offered a compromise. Garuda would become immortal and would serve as a saddle-animal of the god Vishnu, who stood out as supreme god because of his power. After that Vishnu, 'the penetrating one', 'flew' through the myths on Garuda.

In the autumn I too flew with Garuda. From Bali to Singapore. Garuda is the name of the Indonesian airline. I was told that the Indonesians, aware of the marvellous qualities of the legendary bird, hoped that the name Garuda would give their airline a good name.

I can write these special qualities in the mythological bird's dossier:

—Garuda could fly when intelligently steered.
—Garuda could take on water.
—Garuda could extinguish fires.
—Garuda could put a smokescreen round fire-breathing snakes (laser cannons?).
—Garuda could destroy with 'divine eggs' (bombs?).
—Garuda could fly inside and outside the atmosphere (to the moon).
—Garuda was immune to unknown but powerful weapons.

Very odd.

The bird that carried the Babylonian Etana into space was also described as an eagle. The first manned spaceship to land on the moon was called Eagle!

Is history repeating itself?

A Balinese carving of Garuda.

Who was Shiva?

What is Shiva?

The answers to both questions reveal a mysterious background.

Shiva was one of the main gods; he is described in detail in the Indian Vedas. He had his permanent dwelling on Mount Kailasa in the Himalayas. His name in Sanskrit means 'the kindly one', 'the friendly one'. These qualities must have prevailed, for he was also the god of destruction, while being held in high repute as bringer of blessings.

Shiva must have looked rather uncanny. In most pictures and sculptures he appears naked, or clothed in filthy skins, as an ascetic smeared with the ashes of a corpse, with plaited and unkempt hair. In addition Shiva had five faces, four arms and three eyes!

The third eye was in the middle of the forehead. The Vedas relate that he could destroy as well as see with it. If he looked hard at an enemy, a ray of fire shot from this dangerous third eye.

And that's not all. His blue tongue and blue throat had their story. When snake gods poisoned the water, Shiva, with the help of his wife Parvati, managed to filter the undrinkable water through his mouth. From that time on his tongue and throat were blue.

Shiva was considered to be invincible, and kindhearted and gentle when people prayed to him.

Once, the gods, whose head was Indra, were attacked by the Asuras, another ancient Indian group of gods. Although Indra hurled his Vadshra, a dangerous club, at the enemy, he was in such distress that he prayed to Shiva for help. As soon as he heard the prayer, Shiva did not withhold his aid, indeed, he was at once prepared to endow the Indra faction with half his enormous power. Then, he said, they would be capable of destroying the Asuras with a single fiery arrow. But neither Indra nor his companions in arms were in a position to accept and store even half of Shiva's power.

Shiva saw this and proposed that the Indra gods should let him have half their power. They did so and Shiva conquered the Asuras in a flash, but he did not return the borrowed power and from that day was the strongest of all the gods.

Shiva's arsenal of weapons also included the Pinaka, a trident which was reputed to be a flame-thrower. Then there were a sword, a bow and three snakes. These coiled round him and defended vulnerable parts of the body: head, shoulders and loins. It is obvious that the loins needed special care, for his symbol as creator of new life was the phallus, or lingam, home of creative power.

Oscillating between his faculties of production and destruction, Shiva loved the gay and the sad dance, the dance of the 'eternal movement of the universe'. When Shiva himself danced this dance of 'cosmic truth', there was a halo round his head and he was surrounded by ghostly figures.

Shiva, the 'lord of the Universe', could do all this and more. We should take good note of these attributes, but also read between the lines a little.

What is Shiva?

The most powerful laser cannon in the world!

Its home is Livermore, a small suburb of San Francisco. Shiva cost more than its divine predecessor, namely thirty million US dollars! In a milliardth of a second Shiva can fire twenty laser beams at a target the size of a grain of sand. Its energy output: 26 million megawatts. In comparison a nuclear reactor of the normal type produces about 1000 megawatts when operating steadily.

Like the mythological Shiva the modern Shiva can be destroyer *and* saviour. 'Our' Shiva can detonate hydrogen bombs, and also make them explode, *before* they bring disaster. One day 'our' Shiva will be able to solve all energy problems with one blow—by the nuclear fusion of hydrogen with helium. The hydrogen-helium fusion reactor is the dream of all physicists specialising in energy.

What is going on in Livermore?

The Shiva laser beams aim at a microscopically small glass ball. A gaseous mixture of deuterium and tritium, the isotopes of helium, is smelted into the tiny ball. If the concentrated laser discharge hits it, it disintegrates with such incredible intensity that millions of degrees of heat are produced. The point of the experiment: at such high temperatures hydrogen atoms dissolve into helium. The rest is simple, say the scientists. As in previous reactors the energy freed will be turned into steam which will drive turbines.

The 'creators' of the modern Shiva are the scientists of the Lawrence Livermore Laboratory of the University of California. They are convinced that in this way they can solve the energy problem by the end of this millennium. Then, to put it simply, a few litres of water will be enough to supply a whole city with energy. Omnipotent Shiva will make it possible.

History repeats itself.

Tantalus was the son of Zeus and a traitor! Son of a god and King of Phrygia, Tantalus owed his fame to his privilege of being allowed to sit and eat at the gods' table, and listen to their conversations. Instead of keeping silent as befits the confidants of the mighty, Tantalus began to betray the secrets of the immortals to his earthly friends. His image grew and grew. Men looked on him as a chosen spirit who knew far more than they did, as a man who could see behind things and events.

In order to keep in favour with his divine champions, Tantalus offered them a banquet. With the idea of putting the gods' omniscience to the test, he had his son Pelops killed and served at table. Yet before they had tasted a mouthful, the gods sensed his wanton wickedness. They brought Pelops back to life again, but father Tantalus suffered eternal banishment to Hades, the underworld, in which he was to suffer terrible torments. Down there in the dark and wet, wicked papa Tantalus underwent the world-famous punishments connected with his name. He stood up to his neck in clear water, but it ebbed away from him when he tried to drink it; at the edge of the pond hung delicious fruits, but they moved away whenever he tried to grasp them; lastly a rock hung over his head ready to fall and crush him. With these three torments—thirst, hunger and mortal fear—Tantalus paid for betraying the secrets of the gods.

The modern Tantalus is once again about to disclose tremendous secrets—driven to it by research. Scientists at Wisconsin University christened a complicated machine Tantalus. It can accelerate electrons almost to the speed of light. The particles so accelerated emit a strange bluish light which is called synchrotron radiation, a radiation that is much stronger than any dosage of X-rays. It penetrates the structure of molecules and atoms.

In this way the universe will be forced to give up one of its great secrets. The atomic world, previously closed to the human eye, will become visible, the structure of matter will be revealed down to its atomic components. Tantalus the machine is preparing to betray divine secrets. We shall learn to understand the structure of matter and how to form it ourselves. We shall look over the gods' shoulders.

An old story repeats itself.

The guardians of the cosmic secrets have already been forced to yield up a treasure. The Tantalus machine has already been outdone.

The object concerned shone long ago in the story in the Arabian Nights, Aladdin and his wonderful lamp. In this story which is traditional in Europe, Asia and Africa, Aladdin descended into an underground cave at the behest of a magician in order to bring back the wonderful lamp. If anyone rubbed the lamp, his every wish was granted. When Aladdin realised the power of the lamp, he refused to give it up, had all his wishes granted and married a princess, with whom he lived happily ever after.

Aladdin, too, stands godfather to an apparatus that bears his name. Aladdin is being built at considerable expense in the Brookhaven National Laboratory.

Aladdin's rays are to be a hundred times stronger than those of Tantalus from Wisconsin. The new wonderful lamp is intended to reveal the secrets *behind* atomic particles. Here perhaps scientists will find out how to take matter apart, convey it by rays to another place and there reassemble it in its original form. Aladdin fulfilled his dearest wish by conjuring up palaces and people from the void. In the future his successors in Brookhaven may be able to do that, too. The first steps have been taken.

In the distant future it may be possible to move matter without traditional methods of transport. Aircraft, ships, railways and cars will be on the scrapheap. In the TV science-fiction series *Star Trek* trick photography anticipates what may become reality. The film-makers send the crew to the surface of the planet by directional rays.

When Tantalus research is completed, it is hoped that this goal will be reached. Just as a television camera scans pictures and reduces them to many thousands of tiny dots, which are then reassembled on the screen, a cluster of powerful rays will scan solid bodies of any kind, decompose them and carry their molecules and atoms at the speed of light to another place where—presto!—they will be reassembled after their original pattern faster than you can blink. The radiation in Aladdin's wonderful lamp provides the prerequisite. Fairy-tales become reality.

Nearly all ancient civilisations worshipped the sun. In Sumer the sun god Utu handed on the baton to the sun god Shamach, who represented the beneficent powers of the sun. The Egyptians revered their sun god Re (or Ra), whose name was adopted by other gods to show themselves as creators. Amon Re is an example. Starting with Chephren, the fourth dynasty Pharaoh, even the kings called themselves 'Sons of Re'. There were temples of the sun in all the larger towns.

Sun worship was also a characteristic of the Inca religion. The Inca rulers traced their origin to the sun god Inti and called themselves 'sons of the sun'. The ancient Greeks kept their sun god Helios warm by erecting large sanctuaries in his honour on Rhodes. And for simplicity's sake the Romans called their god the same as the Latin name for sun—*sol*.

The stimulus to the global concept of a solar religion was the life-enhancing, creative power of the sun itself. It radiated light and heat, and helped men, animals and plants to flourish. People knew perfectly well that nothing could survive without the sun. Without it the earth would freeze into ice and cold and all life would be extinguished.

In this Year of Our Lord 1979 *we* stand at the gateway to a new global solar religion. Once again the sun will be celebrated in 'sanctuaries' and take us all in its spell and promise hope. All over the world solar energy will replace atomic energy, which is so unjustly calumnied.

The first 'Temple of the Sun' has been working for ten years near the village of Odeillo in the French Pyrenees. In consistently fine, sunny weather, it can produce temperatures of up to 3000° Centigrade. Large US firms such as Boeing, McDonnell Douglas and Exxon, German concerns such as Dornier and Messerschmidt-Bölkow, Swiss enterprises like BBC or the Israeli Holon Institute compete with the Russians to achieve the most successful solar technology. At the end of this intensive research we shall capture x thousand megawatts of solar energy via gigantic satellites and feed it into the new 'Great Temples', the solar power-stations, by microwave. It is intended that industrial layouts and all appliances in daily use will rely on solar energy in the not too distant future.

Does this mean that there are no dangers in the use of solar energy? Of course there are, for no form of energy can be handled without a certain amount of risk. The modern sun-worshippers do not talk about it, because it does not suit their book. First they have to damn the cleanest energy, that produced by atomic power, on ideological, not factual grounds. In other words the kindly old sun is shining on our roof.

But supposing that for some reason a micro-wave transmission did not hit the gigantic receiver antennae on the earth, the effect would be devastating. Clustered microwaves alter the structure of cells. All organic life—men, animals, plants—would slowly languish and die. And if in case of war the opposing forces reciprocally shot down the solar satellites from the skies, the lights would go out on earth. In the case of global natural catastrophes even the smallest solar energy installations on rooftops would be absolutely worthless.

What kind of natural catastrophes am I thinking of? The new sun temples will be solidly built in places far from where we actually use them—in desert areas, for example, on the sunny slopes of mountains or as floating solar cells on the seven seas. Great distances have to be covered from these receiving stations to the centres of civilisation where the energy will be used. If the earth's axis changes its position only minimally, earthquakes and floods of biblical magnitude will occur. If the supply routes are interrupted, mankind will lack vitally needed energy just when it is wanted most.

Climatic changes on a vast scale are undoubtedly taking place. Possibly they will so alter the global weather outlook that regions which were *once* controlled by areas of high pressure, and so were preferred sites for modern solar temples, will revert to areas of low pressure with predominantly dense cloud cover. Then our temple of the sun will be in the wrong place! Scientists will warn of years of violent eruptions on the sun and sun spots, which may cause and speed up climatic changes.

Never mind. The sun gods Utu and Shamach, Helios and Re and Sol, and all the others, are revered again. Once again people pray to the life-giving sun. The temples which the modern 'believers' build are incomparably dearer than the ancient ones whose stark ascetic ruins impress us. We are determined to worship the sun. Nothing functions without the trademark Helios. As in earlier times. History repeats itself.

Affectionately, almost romantically, twentieth-century technologists give mythological names to their forward pointing discoveries.

Why?

Is mythology overtaking us?

Are we in the process of turning mythology into reality?

Only a few years ago anyone who claimed that divine lightning was more than a mythological name would have been ridiculed. I am surprised that no mythological name was found to fit its earthly counterpart. The invention must have been too surprising. Laser! Everyone knows the name today, but few of us know what it stands for. *L*ight *a*mplification by *s*timulated *e*mission of *r*adiation.

Last year, 1978, the Americas alone expended 500 million US dollars on the development of laser cannons. If we could add up the investments made all over the world, the sums would be truly astronomical.

Why this lavish expenditure? Divine lightning is meant to explode rockets in flight, to take 'wicked' satellites out of the sky. Everything that was once constructed by the earth to use in space may one day have a 'reverse' significance. Cities on earth may be 'evaporated' from space platforms. The divine lightning of mythology has suddenly become reality. History repeats itself. . . .

But laser is also suited to peaceful activities.

On 24 June, 1978, an uncanny sight could be seen in the sky above Atlantic City. Without exploding rockets, without crackling phosphorus wheels, a fantastic play of light lit up the sky. Heinz R. Gisel, international promoter of celestial light orgies from Zurich, says that his enterprise can 'paint' whole scenes in the sky with laser, phantasmagoria visible over 20 km away.

There is no doubt that we are only at the beginning. Ultimately there will simply be no limits. Pictures and writing will be projected into space for 1000 km. On the side of the new moon facing us, flickering letters will read: Coca Cola—Coca Cola—Coca Cola. At full moon religious communities will warn us with black laser beans: the Last Judgement is at hand!

Now I have nothing against this technical magic. If I live to see it, I shall even smirk, because I know that all these things, which for us still lie in the lap of the future, have all existed before. Gods, who appeared with 'great power and splendour', projected pictures into the firmament and our technologically unskilled ancestors did not understand them and took them as signs of divine power. Sometimes they saw hands writing, sometimes faces or—as described in the Old Testament—a pillar of fire which shone at night and even by day looked brighter than the light of the sun.

Our space technologists have taken a liking to mythology. They call their satellites Midas, Samos, Cosmos, Pegasus, Helios and so on. They gave their heaven-storming rockets names like Thor, Atlas, Titan, Centaur, Zeus, Jupiter, Saturn and Apollo. Almost the whole great mythological family is assembled in the sheds of the space railway stations. Nostalgia for the past or a prospect of the future?

In the year AD 5000 (if so-called historical time is still being counted from Christ's birth in those days), etymologists will be struggling to explain this text:

> The first sign came from Helios. The High Priest assembled the wise men of the land and informed them: the godless enemy is trying to destroy us.
>
> The Council of Wise Men decided that Saturn, the mighty, should take Samos to heaven to get information on the spot and report his observations to the priests.
>
> Hit by a lightning flash of incredible power, Samos fell into the depths of the ocean. There Neptune, the god of the sea, spoke: 'Transfer a third of your power to me, then I shall ask Nautilus, who dwells under the eternal ice, to help us. The agile children of Zeus are with him.'

Nautilus rose noiselessly from the water, filled with Neptune spirit. He broke through the ice-covering below the North Star and sent the children of Zeus to invade enemy territory on glowing rays. Night became bright as day. It was horrible to see how men were convulsed and crumbled to dust.

While the hostile territory was burning up in the terrible heat, the heavenly guardian Pegasus reported that the godless were entreating Shiva, the destroyer of all, to help them.

When the high priest learnt this, he realised at once that only Nora in the full bloom of her power could be effective against Shiva. Equipped with the fascinating gaze of Helios and Pegasus, he ordered the assembly to concentrate their prayers on Nora. Nora's veins were filled with power. The habitations of men grew cold. All the angels stood still. With a low humming sound, the children of the gods transferred their energy to Nora.

Then a glowing bright lightning flash flickered forth, lighting up the earth more radiantly than the sun; it rose into the farthest heavens. The deathdealing eyes of Shiva were blinded. The powerful mighty god tumbled helplessly on to the moon.

Even in 5000 years the mythological names will still cause confusion and appear mysterious. Yet the *en clair* message is so simple:

The satellite Helios reports that the enemy has taken up an attacking position. The supreme commander calls his staff together and confers about the situation.

It is decided to put a Saturn rocket, supported by some Samos satellites, into orbit round the earth.

During the reconnaissance mission Saturn is hit by a laser beam.

The commander of the naval forces takes the view that he can start a counter-attack from the atomic submarine Nautilus provided it manages to divert the enemy forces from its position by fake attacks. Nautilus is equipped with 20 Zeus rockets.

Programmed by the computer Neptune, the supreme commander, Nautilus breaks through the ice under the polar cap and fires its rockets at the target. The enemy commando replies with laser beams, energy for which is drawn directly from the sun. The territory burns up.

Satellites of the Pegasus type measure the energy charge and radio the results to headquarters on earth.

There they recognise the danger at once. There is only one laser which is stronger than the one on the platform in satellite orbit, the laser of system Nora. Nora's laser cannons are fed by an atomic reactor which is not yet ready to operate. Consequently all the energy in the country is concentrated on Nora. Machines in the factories come to a halt. Houses have no light.

With this concentrated power Nora succeeds in hurling Shiva out of orbit round the earth. Hit by the enormous power of the beam, the platform crashes on the surface of the moon.

We only have to imagine a man of the pretechnological age rising from the tomb and being confronted with our own technical achievements. He would see enormous aircraft rising into the sky, moving pictures flickering in one corner of the room, etc., etc. As this man who was resurrected would have no explanation for the 'phenomena', he would think they were magic.

We face the 'puzzles' that mythology hands down to us on a reverse footing. Looked at in the usual way, everything has developed nicely and continuously, everything began on a small and primitive scale to become sophisticated and complicated later. There is obviously no place for technology in the distant past in this way of thinking. So if accounts of 'flying machines', ray guns and destructive weapons crop up, it can only be a matter of fantasy, magic or ideograms! Excuse my mirth!

I read a newspaper report that the USA is looking for a cloak of invisibility for its aircraft-carriers on the high seas.

Cloak of invisibility? I have heard about that before. A cloak that made one invisible was always a sought-after requisite. The ancient Germans firmly believed that elves could make themselves invisible by donning a cloak. In the *Nibelungenlied* the hero Siegfried won the cloak from the dwarf king Alberich and used it afterwards in successful duels.

But there is a considerable difference between making a person disappear and a gigantic ship of 100,000 tons displacement which travels at 70 kilometres per hour.

Is the US Navy on a hopeless search?

For a long time now enemy attacks on a target have not used a telescope to spot the objective. The human eye is not a reliable operator in mist and rain. Targets are picked up by an electronic 'eye'. There are two ways of getting the target in the sights. The attacking rocket either flies to the target according to the programme it has been fed with or it finds the target quite independently by rays which it emits: radar, infrared- or microwaves. Target seeking is always done automatically, using target data programmed by a computer or electronically calculated in fractions of a second. If the electronic system can be misled, the technological cloak of invisibility has been found. Is the deception practicable? The newspaper *Die Welt* comments:

> 'As the Hughes Company has recently informed us, the attacking weapons are so deluded electronically that they attack a "ghost target" of the dimensions of an aircraft-carrier and literally fall into the water. All we can conclude from the hitherto scanty data is that antennae are installed on both sides of the aircraft-carrier. Mini-computers are fitted in the operations room. The system is supposed to be able to follow several hundred signals simultaneously.
>
> On receipt of a signal the system automatically measures the transmitting frequency. The computer separates friend from foe. Then it selects the tailor-made electronic 'cloak of invisibility' for the specific frequency spectrum. The enemy flies to an electronically depicted but

non-existent target.' (17.5.1978)

Fantastic, but that, too, has existed before.

In the Indian national epics *Mahabharata* and *Ramayana* we can read about weapons and flying objects which could make themselves 'invisible to the enemy'. Even the ancient destroyer Shiva sometimes disappeared into thin air before the eyes of the enemy. History repeats itself. That is a model worth emulating. We are on the way back.

Reports like the one that NATO possesses an antitank rocket that hits the target regardless of the weather is no newer than the invention of the neutron bomb which only kills living creatures and leaves inorganic matter undamaged. Water under the bridge are also crossings of men and animals, of man and machine in cybernetic units, whose practical application will be announced for the next century. But there is a little more to be said about that.

Is history repeating itself in a fatal way?

For years a really fatal 'example' has been worrying me. As I told you that I wanted to introduce dangerous ideas into the conversation, I can let it out at last.

Let us take the terrifying idea that crazy minorities unleash a global war of destruction on our planet.

What targets would they turn their murderous weapons on?

On the uninhabited Sahara? Surely not.

On the inaccessible Himalayan mountains? Hardly.

On the ice caps at the North and South Poles? Why?

On the settlements of poor Indians in the South American Andes? Never.

On the palm-clad atolls of the South Seas? What for?

On the haunts of the aborigines in Australia? Never.

On the huts of the negroes of Central Africa or the poorest of blacks in the so-called Republic of Mali? What would be the point?

On the North American Indians in the deserts of Mexico and Arizona? Definitely not.

On the descendants of the Maya in the jungles of Yucatan? Scarcely.

On the jovial Russian farmers in the wilds of the tundra? Without rhyme or reason.

On tribes in the Amazon? What harm have they ever done?

The targets of the warring parties will certainly lie in the centres of civilisation where millions live and work. These are precisely the territories that are supposed to vanish from the map.

Now it is not true that massive atomic attacks will subject our planet to radioactivity for all eternity; life will continue, especially in those places where no bombs fall. Moreover living creatures, including man, are more adaptable than we think. On top of that, modern and future weapon development tends to evolve destructive weapons with 'cleaner' radioactivity. Limited territories will be saturated with deadly atomic bombs, but they will only be effective for a limited time. Both attackers and defenders are equally interested in having such weapons. What use is victory over a country that can never be inhabitable or cultivable again? Who gets anything out of a radioactive Europe which the victor cannot enter?

At all events men, groups of men, will survive in the Sahara, in Tibet, at the Poles and in the Andes, in the South Seas and the interior of Australia, in Africa and the deserts of Mexico, in the Russian tundra and the Indian reserves of Yucatan and the Amazon—but there will also be survivors of the catastrophe in the civilised nations with advanced technologies.

There may be thousands or hundreds of thousands of people who survive the global conflagration—they live scattered throughout the world. They know nothing about each other. They all wish and hope that they are not the only survivors, yet there are no contacts or news. Each and all of them are an island.

The survivors speak different languages and dialects. If they did make contact, how on earth would they understand each other? Radio, television, telex all means of communication have been destroyed. It is like day zero. No factories are working. There is no supermarket with goods for sale. No cars on the streets. No aircraft in the sky. The survivors are on their own. The big Robinson Crusoe adventure begins.

In those days a western engineer succumbed to the enticements of a travel agency and had begun his holiday in the highlands of Tibet when the great war broke out. Familiar with all the horrors of atomic war, this man knows that no kind of transport will ever take him home and he also knows that he no longer has a home. What does he do?

He has been fully trained as a technologist and is therefore far superior to the Tibetans in technical matters. Like the ancient Greek mathematician and inventor Archimedes (285–212 BC) he can make his discoveries anew, discover his law of levers again, recalculate the content of areas and bodies and the Archimedean point outside the earth again. The Tibetans admire him.

On the basis of his knowledge, the engineer calculates that somewhere on the earth other groups, too, have survived the atomic attack. In any case he wants to know exactly what happened elsewhere. Curiosity spurs him on . . . as it does other groups who survived and have intelligent technologists in their ranks. Sooner or later our engineer will go on an expedition.

Other groups will do the same. Each group suspects that there will be fellow-men who have survived.

Before they set out they will leave messages behind in case strangers visit the place they are vacating. What language would they use for the messages? They would have to say succinctly and intelligibly:

—We were here and we are coming back.
—There is drinking water here.
—We are going north (south, east, west).

—We are led by an engineer (priest, architect, pilot, etc.).
—Danger: deadly small animals.
—Warning: a very aggressive native tribe 40 miles to the north-west.
—We possess the pre-catastrophe range of knowledge.
—A doctor lives in the north gorge of the mountain.
—Berries are poisonous. Do not eat.
—Edible fish in all ponds.
—Contaminated territory to the north and west!

Living in need themselves, they will make a present of all their experience to other survivors, their brothers in need. They will deliberate how to explain why they are going where, whether women and children are with them, whether mutants immune to radioactivity are on the expedition.

The original question remains: in what language do they tell the strangers?

The Tibetans do not speak a word of English, and even our engineer is baffled when a Russian addresses him, while Cyrillic characters are like surrealist drawings to him. What is the answer?

Tourism brought millions from all countries to other parts of the world. International athletic meetings mixed people from many countries together; they too, were victims of the catastrophe. Is it not reasonable to assume that educated intelligent men were among the survivors everywhere? Saharan Arabs do not understand the South Sea islanders. Should they speak English? Russian? Chinese? German? Or French, the language of diplomacy? Or any of the other 3900 languages?

There is only one international means of communication—*picture language*. And that goes for modern man, too.

That has been tried and proven daily. The Indian who lands at Frankfurt finds his way through the confusion of the airport because pictures show him the way. To the exit, the luggage collection hall, the customs, the WC's, telephones and taxis. In the spa of Baden-Baden the non-German-speaking Australian sees at a glance where there is spa water, where the theatre is, where to find the swimming-pool or an emergency doctor, and what sights he should not miss. At the Olympic Games everyone is told by picture where to change money, where interpreters can give additional help, where the bicycle-racing track is and where the outdoor orchestra is playing.

All without speaking or writing!

By pictograms.

The last twenty years have seen the development of more than 500 universally intelligible pictograms which enable even the illiterate to visit foreign countries without being able to read. In its travel brochure, Baden-Baden alone includes more than 100 pictograms used in the spa that have long since become reliable tourist guides. The pictograms have achieved what Esperanto never did: communication between people without the need for speech.

Pictograms can be more than simple indications. They can be used to form whole sentences. For example, a bunch of grapes = wine, a man with a castle in the background = this is the way to the castle, a man aiming a gun = this is shooting land. The foreigner gets a clear message from the three pictograms: 'If you want to drink a glass (or more) of wine, please take this road to the castle; you can also shoot up there (if you have a permit).'

A mathematician could tell us how many combinations can be made out of 500 pictograms. It is certainly a figure far higher than one's chances of winning the football pools.

Pictograms are the international language of our time!

Back to our groups which survived the great catastrophe. Even if they were not familiar with pictograms at home, they would have to invent them in their need. Every intelligent being would realise that it was pointless to write messages in 'his' language. The obvious thing is to think out simple stylised figures and symbols and chisel, scratch or scrape them into rock-faces. Signs that they themselves would understand if they met them.

Above: modern pictograms in the streets of Baden-Baden.
Below: ancient pictograms from rockfaces in British Columbia and California.

For twelve years I have been travelling like a tramp all over the world. Among the Hopi Indians, USA, in the ghost town of Sete Cidades, Brasil, in Kashmir and Turkey, in South Africa and the Sahara, in Northern Europe and Southern France, in California and North Italy, in the South Sea islands and in the Philippines I have photographed rock drawings and engravings. I got to know White Bear, a chief of the Hopi Indians. He took me to a secret canyon on the reservation which the Indians protect from the curiosity of strangers. The walls were covered with 'pictograms'. I asked White Bear if he could read the signs. Not all, he said, but he did know most of them

I want to know for whom and for what purpose his ancestors had left these signs.

The old Indian explained to me that his ancestors had emigrated from south to north—not, as scholars claim, from the Bering Straits in the north to the south—and that the tribe had often split up and formed new groups during the great trek. In order to convey the experiences of the groups who had gone ahead to those following behind they had used rock drawings.

Why, it flashed through my mind, are there rock drawings from different epochs?

White Bear knew the answer. Different groups and their descendants returned to the same places to carve new discoveries and good and bad news. Rock drawings and engravings had the same value as wall posters have for Mao's Chinese today.

Oswald O. Tobisch collected some 6,000 rock drawings and compared them with each other. Using twenty tables he demonstrates how closely European, Asiatic and American picture-writing groups are connected. Tobisch concludes from his comparative study that all cultures *must* have been reciprocally influenced, indeed that ultimately the rock drawings must have had a common origin (34).

The Indians still use pictograms today. They have never

Pictograms of the kind Red Indians still weave into their typical tapestries.

stopped using ancient stylised patterns, i.e. pictograms, in their typical artistic products. The subjects of the North American Indians' sand drawings 'tell stories' in the same way and even the tapestries of the Andean Indians include pictograms.

Can the phenomenon of the millions (!) of rock drawings all over the world be explained by a world-wide catastrophe?

Would history repeat itself after a catastrophe in the present?

Would survivors seek the path into the future, their union with other survivors, by means of rock pictograms?

Is the past approaching us; is it overtaking us?

Is the present seeking the 'kiss of death' with past history?

If the latest weapons systems are christened with mythological names, if we rediscover an internationally intelligible picture writing, if our delving into the deep dark springs of the past are so obvious, does *the original reason* for this lie in antiquity, in early history, or is the reason inherent in ourselves?

Is our consciousness a *perpetuum mobile*, an eternal cycle, the paths of which lead from past into the future, from the future into the past? Where does it begin, where is its primal cause, where is its original impulse?

Is it presumptuous even to pose the problem of the original spark that set off this cycle?

Arnold Sommerfeld (1868–1951) would be assured of his special status in the sciences simply because three of his students won the Nobel Prize: Werner Heisenberg (1932), Petrus Debye (1936) and Wolfgang Pauli (1945). Hans Albrecht Bethe, another student of Sommerfeld's, is one of the leading nuclear physicists and is head of the Department of Theoretical Physics at the Los Alamos Atomic Research Centre.

Teacher of the famous, Sommerfeld himself was a modest figure, although he discovered the majority of the laws governing the number, wavelengths and intensity of spectral lines. His most important book *Atomic Structure and Spectral Lines* was the standard work on atomic physics for decades.

But Sommerfeld was (too) far ahead of his time with one of his discoveries. In addition he had the bad luck to announce it just before the publication of Einstein's theory of relativity, which caused a tremendous sensation in the scientific world.

Sommerfeld put forward the theory that there were particles that were faster than light and had the peculiar quality that the more energy they lost, the faster they travelled.

Einstein's theory obscured Sommerfeld's bold idea, for it claims that particles on the verge of the speed of light have infinitely great mass.

Once introduced to the world, speculative theories with

even a trace of probability do not lose their power of attraction. Since Sommerfeld's announcement at the turn of the century, generations of physicists have 'toyed' with the theory of faster-than-light particles. Gerald Feinberg, Professor of Theoretical Physics at Columbia University, New York, was first able to set the discussion going again in 1967 with his work on faster-than-light particles (35). He also gave particles a name. Feinberg called them tachyons, coming from the Greek word *tachys*=fast. Once again scientists countered that, according to Einstein, nothing could be faster than light, but some experts on elementary particles fell in with the fascinating idea and supported the view that faster-than-light particles must exist.

Can the bold idea be reconciled with Einstein's irrefutable theory?

Einstein's theory of relativity states that a body that does not have the speed of light in *one* inertial system* cannot assume a speed faster than light in *another* inertial system. So if a particle on the borders of the speed of light acquires infinitely great mass, it can neither reach nor cross the boundary of the speed of light.

Is this correct? When it originates and disappears, light itself behaves like its particles, photons and neutrons, which move at the speed of light, indeed their velocity is never less. Elementary particles have already been accelerated to 99.4% of the speed of light, without infinite mass, in every large synchrotron, such as the one in the CERN near Geneva.

What powers the photons and neutrons? What is their 'secret'? All they have is the energy of motion. If they are brought to a halt, they vanish without a trace.

Dietmar Kirch (36) divides elementary particles into three broad classes:

1 Particles like nucleons and electrons. (They travel below the speed of light).

*An inertial system is a non-accelerated frame of reference.

2 Particles like photons and neutrons. (They travel at the speed of light).
3 Tachyons. (They travel faster than light).

First of all, tachyons only exist in an inertial system which is not yet accessible to us. So there can be no contradiction of Einstein's theory. Just as Class 1 particles *always* travel *below* the speed of light and cannot be made to exceed it with finite energy, the tachyons of Class 3 *always* travel *faster* than light and cannot be slowed down to the speed of light.

Tachyons exist in a different inertial system. They behave in exactly the opposite way to the elementary particles in the inertial system which we know and in which we live.

> 'An event can be described by giving the place in space where it happened and the time at which it occurred. Thus an event is a four-dimensional reality. The time datum which characterises an event is not independent of the coordinates which describe its position in space. Because measurements of time and space both alter when the inertial system is changed, we speak of a four-dimensional space-time.
>
> Seen from the point of view of many accelerated systems tachyons can move backwards in time.' (36).

Confusing qualities! Whereas everything moves from the past to the future in our system, tachyons *can* travel from the future into the past.

Can this phenomenon be made intelligible?

Imagine a flashlight apparatus coupled to a receiver that can register tachyons. The flashlight is programmed in such a way that it lights up as soon as a tachyon impulse strikes it. Let us suppose that a satellite emits a tachyon impulse *dead on midnight*. What happens?

It is not yet midnight, but the flashlight lights up *before* the satellite has ever emitted the tachyon impulse. How can the flashlight apparatus accurately programmed for the tachyon impulse react *in advance*?

'Time' in the tachyons' inertial system is not identical with 'time' in our system. 'Seen' from our position, tachyons are travelling *backwards* faster than light. What we in our system know as the causality principle—namely that every effect must have had a cause—is no longer valid as soon as we are concerned with the four-dimensional space-time of faster-than-light particles.

The apparent contradiction would be resolved if we ourselves were in the tachyons' system. Then the physical laws would agree again. In our system the idea of from past to future is 'logical'. *We* cannot imagine the cause coming after an effect. If there are intelligent beings in a tachyon world, presumably *they* would not be able to imagine why the future always had to follow a past. For them a much more normal process would be the deduction of the past from the future. If we speak of the 'distant past', they speak of the 'distant future' in the tachyon world. But it means the exact opposite of what we think of as the future.

Now we must ask ourselves: What is time? What is past? What is future?

In our consciousness 'time' is the passing of the present; the past grows.

This simple definition is no longer accurate as it has been proved experimentally that every inertial system has its own inertial time. Even in the choice of identical normal watches times are different in different systems. Scientists all agree that 'time' can only be defined in relation to an inertial system. And as no inertial system seems to be superior by the laws of nature, it is physically meaningless to speak of '*the* time'.

We must change our way of thinking. If an event can occur *before* a cause, what are our points of reference?

The human brain functions in chemical and electrical ways. It develops the imponderables 'mind' and 'consciousness' which are not conceivable or measurable in physical terms. Telepathic experiments have clearly proved that 'consciousness' both transmits and receives waves. 'Consciousness' is also capable of precognition, as parapsychological research calls this faculty. It seems as if 'mind' and 'consciousness' are timeless, as if an unknown form of energy entered the brain and whispered some future information about which we really should not know. I am not talking about intimations of the future of the kind anyone can have as a result of fear and sorrow. I mean the genuine precognition which is known to parapsychology.

What actually happens in our brains? Should we imagine some subatomic particles from another dimension, from another inertial system, that provide our consciousness with information about the future? Have events which took place in the distant past already happened in the future? Do we think in a two-sided canal in which past and future information flow into each other? Is it no coincidence, is it right outside our free will, when we christen new present-day technical arrangements by mythological names?

If time is manipulable in the future and the past, where does the direct effect of time stand? It is a grotesque idea, but could we travel into the past with a hypothetical tachyon time-machine and make an event which took place in the *present* retrogressive? To give an example, could anyone travel back to the ancient Roman Empire on the tachyon time-machine and warn Julius Caesar of his imminent assassination in the Senate? Would the emperor attend the debate in the Senate regardless and be stabbed, as did happen, or would he stay away and change the course of history completely? Shall we be able to influence the distant past from the distant future? Will our descendants in the year AD 10,000 already dominate this kind of 'manipulation'? From this fantastic point of view, is history still unalterable, because it has already been 'corrected' from the future and must unfold exactly in such and such a way and no other for reasons inscrutable to us?

If the technology of spacetravel manages to achieve flights close to the speed of light in fifty years—and it will, provided the Black Order of Pessimists does not succeed in wrecking our future—will spacetravel be a first time event for mankind, or shall we only be repeating what our forefathers already did? Am I not contradicting myself with the sentence 'Our forefathers practised spacetravel', since I firmly assert that *extraterrestrials* visited early mankind?

It may sound dogmatic, but I am not contradicting myself. Perhaps this example will clear up the apparent contradiction:

Let us assume that there was a highly technological industrial society on earth 50,000 years ago. Let us also imagine that our technically advanced ancestors had sent high-speed spaceships to another solar system. During these journeys the terrestrial spacetravellers were subject to the laws of time dilation. As the size of the differences in time depends on the speed of the spaceship, we can imagine that 40,000 years went by on earth as opposed to a mere ten years on board the ship.

Now let us speculate that terrestrial civilisation was wiped out in the 40,000 years between 50,000 and 10,000 BC. By terrible wars, by natural catastrophes, say the breaking up of a polar region and the allied world-wide inundations. By a cosmic event, such as the invasion of bacteria from outer space.

The survivors would have to start again from scratch. Generations after the knock-out blow men are still living in caves. They can write, make fire, fashion tools and social communities, but they only know the glorious past of their own race from the traditions of their forefathers.

Then the crews of the spaceships sent out in 50,000 BC burst in on this fresh start. The crews have only changed by ten years.

What will the spacetravellers do? Save what there is to save. Thanks to their superior knowledge, they will rule the survivors and introduce them to the old laws and rules of communal life.

In other words, our ancestors are visited by their own

ancestors who come from outer space. Here, too, they would be 'gods', who came from the depths of the cosmos, even if they are descendants of *one* family. History repeats itself.

Am I speaking about past or future, when I suspect that in the not too distant future a spaceship with room for a mixed crew of men and women will be fitted out somewhere in the world? Cultures of various bacteria will be kept in its sterilised frigidaire. Vacuum-packed plastic containers with seeds of all species of plant will be stored in its cargo holds. Little fish will swim in oxygen impregnated basins. The study cabins will contain encyclopaedias with the complete knowledge of our time and there will be shelves full of microfilms of all our technical and scientific knowledge. In its work-rooms simple impliments—shovels, rakes, tents—will help the crew to have a chance of survival even at the remotest end of the universe.

A day will come when the crew will tick off all the necessary items on a checklist . . . and set off for the heavens.

And as history repeats itself, the captain of the spaceship will probably be called Noah.

Communiqué

Years ago a friend pointed out that the British Museum housed pictures of tanks that had been used in battles in Sumeria and Babylonia.

On my next visit to London I was able to verify that the ground floor of the British Museum actually does have large reliefs from Babylonian and Assyrian times that show tank-like vehicles. According to archeologists they are representations of battering-rams of the kind used to break through town walls.

This may be, but is not necessarily so.

Four things struck me about these 'battering-rams':

—Battering-rams, whatever one imagines by the word, were manned by soldiers. They do not move of their own accord, and certainly not uphill. Even if the crew were protected against arrows and stones, their feet must have been visible. Even battering-rams have to be moved in some way or other and we can see wheels on them. So how were they driven?

—The ram at the front of the 'battering-ram' can only have an effect if it hits the wall or tower to be stormed at right angles. Rams pointing upwards, as clearly shown in the pictures, make no sense. Kinetic energy has no effect. The upward-pointing ram would have shattered the machine itself on the rebound or sent it up like a rearing horse.

—The twin rams to be seen in one picture are quite pointless. If two pointed rams were banged into the wall, their destructive effect would have been reduced by half. But the builders must have been really simple to make both rams point upwards as well.

—Last but not least why does a 'battering-ram' need a tower?

These two examples of 'battering-rams'—there are several—made me wonder if they were sonic cannons of the kind used at the storming of the ancient city of Jericho.

'. . . when people heard the sound of the trumpet . . . the wall fell down flat, so that the people went up into the city, every man straight before him. . . .' Joshua, 6, 20

'Tanks' in antiquity? Two examples from the British Museum.

5: Signs of the Gods? Signs for the Gods?

It happened in Athens a few years ago. During a press conference, I noticed a grey-haired man, who asked no questions, but was busy taking notes. Afterwards he came up to me and asked very politely if I knew that all ancient Greek temples, including those which date to mythological periods, stood in exact geometrical relationship to each other.

I must have pulled a face, for he assured me that it was perfectly true. But I am well aware that my listeners like to make me happy by giving me pointers that may lead me to new speculations in my field. No, I replied, I knew nothing about it and I thought it was nonsense, because I could not imagine that the 'ancient Greeks' had the geodetic knowledge to fit temple layouts into a geometrical pattern. Besides, I said, the temples were often hundreds of kilometres apart. Sometimes mountains lay between them, blocking the view from one building to another, while he ought to remember that some temples were on small islands which were barely visible from the mainland with the naked eye. No, I summed up, I could not imagine what reason the builders could have had for bringing temples and religious sanctuaries into a geometrical relationship.

The man shrugged his shoulders apologetically and left. My scepticism had disappointed him and I soon forgot him. But I suddenly remembered him when two serious books confirming the Greek gentleman's claims landed on my desk. One was by Dr Theophanis M. Manias (37), a brigadier in the Greek Air Force, the other by Professor Dr Fritz Rogowski (38) of the Carolo-Wilhelmina Technical University at Braunschweig. Both authors clearly prove that *all* the religious sites, oracles, for example, and all the temples of ancient Greece were laid out according to a 'geometric-cum-geodetic triangulation pattern'. After reading both books I remembered my conversation in Athens. I should like to apologise to the gentleman for my nonchalant scepticism, but I don't even know his name. However, he will know about my volte-face when this book is published by Notos in Athens.

The simple fact of buildings laid out according to geometrical principles should not be a 'miracle', for ancient Greece produced one of the greatest mathematicians of all time—Euclid, who lived towards the end of the fourth century BC, taught at the Platonic University in Alexandria and in his fifteen books covered the whole spectrum of mathematics, especially geometry. Was it his idea to place the buildings as they are placed?

Euclid was a contemporary of Plato, the philosopher, who was also active as a politician. Plato sat at Euclid's feet in Megara and listened to his lectures. Was Plato fascinated by his colleague's ideas? Did he put his knowledge to use when he had to share political decisions about building commissions? Was that how the architects were instructed to build the temples in a triangulation system?

Unfortunately this hypotheses is wrong, because most temples and holy sites existed long *before* Euclid!

Nevertheless, Plato must have known about the mysterious geometrical network of ancient Greek buildings, for he mentions a whole series of geometrical contexts in Chapters 7 and 8 of his *Timaeus*. Plato, master of the limpid dialogue, held geometry in high esteem. Today many books on geometry are still introduced by Plato's sentence:

> 'Let none who are ignorant of geometry have a say. Geometry is knowledge of the eternal being.'

It is quite possible that Euclid told Plato about his observations of *already extant* geometrical puzzles. But in that case Euclid must have had access to the primordial geometrical knowledge which became stone in the temples and sanctuaries of ancient Hellas. Dr Manias does in fact say: 'The whole of Euclidean geometry comes from an age-old religious and scientific codex.' (37)

Of course we all know what the 'golden section' is—even Euclid wrote about it. But before I give some staggering examples of geometrical relations between religious sites laid out on the principles of the golden section, I should like to quote the definition which I took from my daughter's textbook (39):

If a line AB is divided by a point E so that the whole line is to the longest section as that is to the shorter section, then the line AB is said to be divided into the 'golden section'

E

A B

If we extend a line divided into the golden section by its longer segment, the new line is divided into the golden section again by the end of the original line, B. This process can be repeated indefinitely.

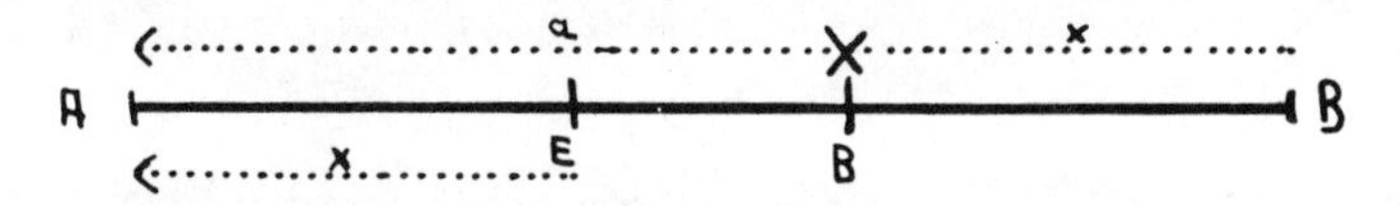

Now some samples:

—The distance between Delphi and Epidaurus corresponds to the longer segment of the golden section of the distance Epidaurus and Delos, namely 62 per cent.

—The distance between Olympia and Chalkis corresponds to the longer segment of the golden section of the distance between Olympia and Delos, namely 62 per cent.

—The distance between Delphi and Thebes corresponds to the longer segement of the golden section of the distance between Delphi and Athens, namely 62 per cent.

—The distance from Sparta to Olympia corresponds to the

longer segment of the golden section of the distance from Sparta to Athens, namely 62 per cent.

The distance from Eipaurus to Sparta corresponds to the longer segment of the golden section of the distance from Epidaurus to Olympia, namely 62 per cent.

—The distance from Delos to Eleusis corresponds to the longer segment of the golden section of the distance from Delos to Delphi, namely 62 per cent.

—The distance from Knossos to Delos corresponds to the longer segment of the golden section of the distance from Knossos to Chalkis, namely 62 per cent.

—The distance from Delphi to Dodoni corresponds to the longer segment of the golden section of the distance from Delphi to Athens, namely 62 per cent.

The geometrical curiosities are not exhausted with the arrangement of the religious sites in the golden section.

If we describe a circle the centre of which is one holy place and which runs through a second religious site, the

Epidaurus – example of an ancient religious site related to Delphi and Delos by the Golden Section.

circle always touches a third and often a fourth site. For example:

—Centre Knossos. Sparta and Epidaurus are on the circumference.

—Centre Taros. Knossos and Chalkis are on the circumference.

—Centre Delos. Thebes and Ismir are on the circumference.

—Delphi, Olympia and Athens are equi-distant from Argos.

—Sparta, Eleusis and the oracle of Trofonion are equidistant from Mykene (37).

Dr Manias also discovered that any temple or site taken as a point also lies on a straight line through two other holy sites.

The incredible thing is that most of these geometrical relations go back to much earlier days than the lifetimes of Pythagoras (about 570 BC) and Euclid, the two mathematical geniuses. In fact they go back to the mythological Greek *Stone Age*. Brigadier Manias shows that seen from a great height the arrangements of the sanctuaries reveal enormous circles, regular pentagons, five-rayed Pythagorean stars, pyramids and even geometrical figures from Greek mythology. To take only one example: According to legend Apollo turned himself into a dolphin and showed the site of Delphi to the priests of Crete. If you draw lines connecting the religious sites between Crete and Delphi, they form a dolphin over 500 km long!

The whole thing is most confusing. The countless geometrical regularities quite exclude chance as the master-builder.

So how can we explain the mathematical perfectionism? How can we reconcile it with the standard of mathematical knowledge we attribute to prehistoric peoples? How did they know at what precise point they had to build?

As the complicated relationships are only recognisable from a great height, we must ask whether 'someone' instructed the builders, whether someone worked out a geometrical network of sites all over Hellas, sticking flags in the ground and saying: This is where you must build a temple!

Or did the ancient Greeks—as Professor Rogovski (38) suggests—first work on a very small scale which only later developed into the large geometrical network? If that were so, why did Plato expressly say in his *Timaeus* that the geometrical relationships were *sacred knowledge which had been handed down for several thousand years*? And if he spoke of 'several thousand years' around 400 BC, we are right in the middle of the age of the gods.

Such puzzles always give rise to a series of similar questions. If we assume that temples and religious sites were built *before* Euclid and fitted into the schema of the geometrical pattern, we must ask why did the Greeks build in that way. We must find out the reason for this extraordinary kind of planning. We must also explain where they acquired such vast mathematical knowledge at such an early date. Lastly, it would be interesting to know who showed the Greek tribes the way to the locations if they themselves could not find them. The series of questions puts us in a dilemma.

But the affair gets even more confusing.

To his surprise, Dr Manias has found that the geometrical system of the ancient peoples is not confined to Greece. The temples of Cyprus, the Lebanon (Baalbek), Alexandria, and even the Egyptian pyramids are included in the network.

The Russian researchers Goncharov, Makarov and Morosov worked on a map of the world which would include all the important ancient cultural centres. When Nikolai Goncharov of the Moscow University of Art saw the finished work, he could not help thinking that he was looking at a picture of a football. (40) The points marking important ancient cultural sites described a ball with twelve pentagonal panels on the globe. Nikolai Bodnaruk, correspondent of the *Komsomolskaya Pravda*, wrote:

'According to the map many ancient cultures did not inhabit chance locations or areas, but were settled precisely on the focal points of this system. This was true of the Indus culture of Mohenjo Daro, of Egypt and North Mongolia, of Ireland and Easter Island, of Peru and Kiev, the "Mother of Russian cities".

The oil-bearing regions of North Africa and the Persian Gulf stretch along the "seam" where the gigantic 'panels' meet. The same thing can be observed in America from California to Texas. Take a closer look at the focal points of the double network: the enormously rich southern part of Africa, the sites of Cerro de Pasco, South America, Alaska and Canada; the underground oil and gas oceans of western Siberia and many others.

Naturally one cannot trace such a connection everywhere. Yet it occurs all too frequently to be the product of pure chance. Besides, deviations from the strictly geometrical pattern are quite understandable, for our planet itself is changing and the formation of natural treasures is still taking place.'

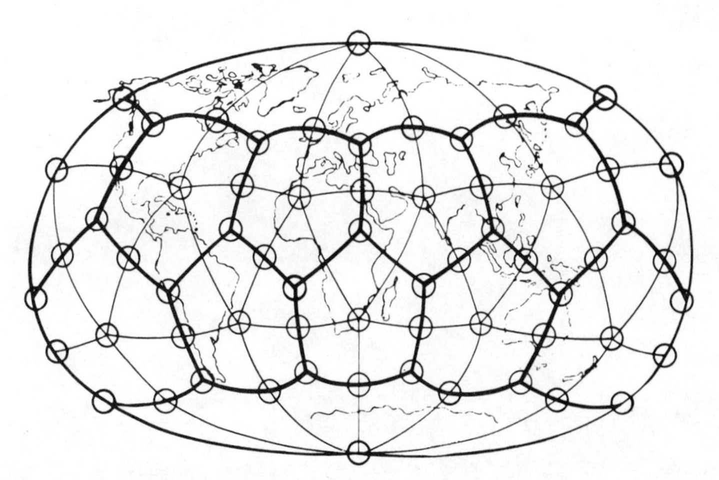

When Russian scholars transferred religious sites from all over the world to a globe, they thought it looked like a football made up of pentagons.

In connection with these recent discoveries, I should point out that Plato had already said in his *Timaeus* that *if we looked at the earth from above, it would resemble a leather ball made up of twelve parts.*

Is there really nothing new in the hills and dales on the face of mother earth?

With my knowledge of the gigantic 'stone signs' all over the globe, I cannot help feeling that the memorials and centres of all cultures were laid out according to the master plan of a global building committee and that signs which could be seen by the flying 'gods' were placed at the 'holy places'. I give these examples off the cuff (more can be found in my 'ancient' books):

—The gigantic pictures cut into the ground between the landing strips on the (now) worldfamous plains of Nazca.

—The gigantic chessboard pattern on the rock faces in the Province of Antofagasta, Chile.

—The 100-metre-high 'robot' in the desert of Taratacar in the north of Chile.

The Maya pyramid of Chichén Itza, a sign for the gods?

—The 110-metre-long White Horse at Uffington in the Berkshire Downs.
—The 55-metre-tall giant of Cerne Abbas.
—The Long Man at Wilmington, Sussex.
—The horse, 13 metres long and 9 metres high, the giantess with outstretched arms, 28 metres high and 21 metres wide, and the 31-metre-tall giant, all three to be seen at Blythe, California.
—The 46-metre-tall giant at Sacaton, Arizona.
—The Boulder Mosaics in White Shell Provincial Park, Manitoba. Silbury Hill, 8 km west of Marlborough, Wiltshire.
—The six enormous octagons with a total length of 11.2 miles, near Poverty Point in Louisiana, USA.
—The Snake Mound, over 400 metres long, at Bush Creek, Ohio, USA. The gigantic concentric circles or wheels to be found at Ripon, Yorkshire, on the Japanese island of Hokkaido near Nonakado and in various states of the USA, not to mention the Medicine Wheel in the Big Horn Mountains, Wyoming.
—Last, but not least, the 250-metre-high Bagpipes of the Andes in the Bay of Paracas, Peru.

This little collection may help to prove that the men of ancient cultures cut gigantic signs on mountain sides or hammered them into plains, although they were only recognisable in all their glory from a considerable height. *Why* did they do it, *for whom* did they do it?

In every case archaeological literature assures us that the signs were associated with an ancient cult. Possibly, but what kind of cult? In all modesty we should like to know that, but we never do. If a cult was involved, it must have been one of the universal sort. It had a common 'denominator' which inspired all peoples to the same activities. How else are the peoples of every continent supposed to have agreed that they must climb mountains and incise signs on them or draw figures on a plain, activities whose end-products were unrecognisable at close hand.

The 'Bagpipe of the Andes', near Pisco in the Bay of Paracas.

The Sioux Indians tell this legend:

'Many moons ago in the past of our ancestors a great wheel came from heaven. It shone like fire and from the wheelnave it twinkled like a star. The winds hissed frighteningly when the wheel came down on Medicine Mountain. The villagers ran away in fear and trembling. When they looked round, from a considerable distance, the wheel rose up—like a wild goose in the bush—and was never seen again. The wise men of the tribe took counsel and decided to surround the spot with stones so that future generations would remember the 'Matatu Wakan', the heavenly wheel, for all time. As the Sioux believed that the wheel came from the sun, they began to etch gigantic signs in the earth that could only be observed by a high-flying eagle.'

We should not laugh at them. 'Cults' like this are still springing up in our sophisticated twentieth century!

The inhabitants of the Melanesian Islands in the South Pacific are tattooed with letters they themselves cannot

Statues of the gods at Tula, Mexico. What kind of gods were these?

read, namely USA. They claim that long ago they were visited by the king of a foreign land called America. The king was called John Frum and he promised that he would return one day from the Masur volcano with 50,000 celestial companions to improve their miserable lives and bring them happiness. However, they say, god John Frum would only return if they observed their customs and worshipped the gods. So they shoulder wooden laths, whisper prayers into primitive wooden boxes from which long palm fronds dangle and perform rhythmical circular dances. What are they up to? They are imitating American soldiers who crash-landed on their island in 1942 and stayed there a while until they were picked up by the US Air Force. This comparatively recent religion was called the Cargo Cult in specialised literature. Cargo means goods loaded on a ship, but the name was given by western cult experts.

On 16 October 1978, the BBC showed a film about rocket launchings in Zaire, Africa. For years the German company OTRAG had been making such experimental launchings in

Mobutu's state to test out a cheap rocket. The camera swung on to a group of negroes who were amazed at the goings-on. An interpreter asked what they thought about it. A negro answered: 'Those are our powerful friends who are sending fire up to heaven!' Who knows if a 'rocket cult' will develop when the OTRAG team have long since departed?

If cults are still initiated in our own day by actual events, we can justifiably assume that cults and myths of the distant past were also inspired by realities, things that really happened. It even makes the gigantic 'stone signs', the signs for the 'gods', plausible. Is that so difficult to understand?

In 1868 the German explorer and ivory dealer Adam Renders got lost in the dense Southern African bush. With his knife he slashed a path through the tropical undergrowth in an attempt to find his way back to civilisation. Suddenly he found himself facing a wall that was ten metres high!

For a moment Renders was convinced that he was safe again, for where there are walls, there are generally men, too. He ran along the walls, but realised that he was going in a circle, as he kept on coming back to his starting-point. Finally he found a hole in the wall covered with brushwood and trees. Renders suspected that he was the first white man to come across the ruins of Zimbabwe.

In 1871 he guided the German geologist Karl Mauch to the spot. Mauch made a plan of the ruins, returned to Germany and claimed that he was the discoverer of Zimbabwe. Mauch supported the theory that Zimbabwe and its environs had once been the site of the dreamland of Ophir from which King Solomon had gold and precious stones sent (I Kings 9, 26 *et seq.*). That was one of the countless explanations which were supposed to solve the mystery of Zimbabwe.

But others located Ophir in India and Elam, in Arabia and East Africa. Probably it lay on the southern part of the

Suddenly Adam Renders came upon a wall ten metres high . . .

west coast of the Red Sea. However that may be, Karl Mauch contributed one of many theories and had no idea that the mysterious place had been reported long before. Incidentally, Adam Renders never left the ruins, staying there until his death.

A dense mist, in which imaginative theories flourish, envelops the ruins of Zimbabwe. The archaeologist Marcel Brion (41) collected all the theories about Zimbabwe and came to the conclusion that they were nothing more than 'romantic speculations'.

It is not surprising that, given its site in the depth of the African jungle, Zimbabwe must have been rather a secret place, for not even the learned Arabian writer and world traveller Abu l-Hasan Mas'udi (about 895), who lived in Baghdad and made extensive expeditions of discovery from there, mentions Zimbabwe in his main book *Gold-washing sites*. However, there is no doubt that large quantities of gold were mined in this area even in Mas'udi's time.

Damiao de Goes (1502–1574), a much-travelled Portuguese historian, does mention Zimbabwe, but he never actually saw it; he was told about its massive architecture by proud negroes. His countryman and colleague Joao de Barros (1496–1570) speaks of Zimbabwe in his four-volumed book *Asia*. He wrote:

> 'The natives call these buildings Zimbabwe, which means "royal residence" . . . No one knows when and by whom they were erected, for the inhabitants of the country cannot write and have no traditional history. However, they claim that the buildings are the work of the devil, because, in view of their own capabilities, they think it impossible that they were the work of human hands . . .'
>
> 200 years later, the Governor of Goa commented:
>
> 'It is reported that in the capital of Monomotopa there is a tower or building of masonry which to all appearances is not the work of the indigenous blacks.'(41)

I was in Zimbabwe, which has long been a popular tourist attraction, in the autumn of 1976. You reach the ruins from Fort Victoria by a narrow asphalt road. Only a few kilometres from Zimbabwe lies the Zimbabwe Ruins Hotel. Several thatched huts form a horseshoe around a shady courtyard. Polite blacks serve you food and drink while you sit at stone tables. Silk bands across their chests announce their job in large letters. Food waiter! Wine waiter! Head waiter! You could live an idyllic life here, if it were not for the almost incessant sound of rifle and machinegun fire from a nearby valley. Mozambique is only one and a half hours away.

In the Rhodesian hotels and guest-houses I got to know, there were black and white waiters, and black and white chambermaids. Black and white taxidrivers form part of the urban scene. There are many whites who do not like the blacks and many blacks who do not like the whites. Is it very different here? Do the Germans like their immigrant Turkish workers? Do we Swiss love the hundreds and

Anyone who has been to Zimbabwe is not surprised that tourists from all over the world are attracted by its secret. Such mysteries act like a magnet.

thousands of southerners who build our motorways and barrages, and cut tunnels through the Alps?

I am not trying to soft pedal the racial problem with these remarks, but it is worth mentioning, because even the ruins of Zimbabwe have been dragged into the political arena. Not so very long ago it was considered shocking in Rhodesia to attribute the buildings to the blacks. And in fact the countless negro tribes in the north and south have not erected such enormous edifices. Organisation and planning were alien to them then, and still are today. Twenty years ago, anyone who claimed that Bantu negroes built Zimbabwe would have made himself very, very unpopular, for political reasons. Negroes were not supposed to be capable of such achievements!

I had a talk with the 35-year-old Rhodesian archaeologist Paul Sinclair of the National Museums and Monuments of Rhodesia, who has been working for the Zimbabwe Museum for many years. On his own initiative he organised excavations in neighbouring valleys and at deep levels found Chinese silks, Arabian pottery, countless Bantu ornaments and strange figurines.

I asked Sinclair:

'Who built the massive buildings in your opinion?'

'The blacks,' he answered. 'In the Shona language Zimbabwe means something like "esteemed" or "revered house". A "revered house" can equally well mean a religious temple or a kind of royal residence. Unfortunately, we have not yet found the grave of the megalomaniac dictator who may have commissioned the gigantic works. So the question of his identity will remain unanswered.'

'What led you to the conviction that the blacks were the builders?'

Sinclair led me to a cupboard with many drawers, which he pulled out one after the other.

'Look, we found all these objects in the Valley of the Ruins. Between here and the ports of Sofala and Quelimae in Mozambique there are about a hundred similar ruins, generally on a more modest scale, but built by the same methods. Granite slabs were split by the application of heat and laid in layers without mortar. In the past the Kingdom of Zimbabwe stretched as far as the Indian Ocean. Presumably the unknown kings of Zimbabwe exported gold in order to obtain other goods from the Arabs and Chinese. Here are the proofs! This is Chinese silk and those are Chinese ceramics which were found in the ground here. We found Arabic cloths, bracelets, fragments of glass and even the odd ornament from India. These finds convinced us that there was a trade route here to the ports on the Indian Ocean (present-day Mozambique). What did they trade with? Gold, of course, for we know that there were gold mines in and around Zimbabwe. The king's title, Monomotata, also points to this, for it means roughly "Master of the Mines".'

'Wouldn't it be more reasonable to assume that the Arabs were the builders?'

'No. The fact that objects of foreign origin were found in much smaller numbers than those which were obviously of negro provenance contradicts that theory. All these drawers are full of finds. Relics of the black construction workers.'

There they lay in the drawers, the figurines which may have been carved around a campfire as a leisure activity. The faces exhibit predominantly negro features, but I also saw some which immediately reminded me of my astronaut gods. They had round heads totally enclosed by a helmet. I rummaged among ivory bracelets, bone necklaces and more refined artefacts of wood, with ivory intarsia.

'If I understand you correctly. Mr Sinclair, the blacks built Zimbabwe, but why and for what purpose?'

The archaeologist thought that Zimbabwe was built as a fortress, a protection against robbery, for even in those days the gold stored there was a much sought after metal.

This answer did not satisfy me at all.

What was it the Portuguese historian wrote, after listening to native traditions?

'They claim that the buildings are the work of the devil because, in view of their own capabilities, they think it impossible that they were the work of human hands . . .'

What does Zimbabwe look like today?

The main feature of the ruins is an elliptical wall, 100 m in length, which encloses an area of some 2000 sq m, in other words an area about the size of two football fields. Today this ellipse is called the 'Royal Residence', which is a rather absurd name, as we know that it is most unlikely that a king ever resided within the walls. No tombs, writing, statues, busts or remains of tools or implements have been found there.

Zimbabwe has no history.

The wall that surrounds the 'residence' is 10 m high, with an average width at the base of 4.50 m. The wall was dry built, without mortar, and used up an estimated 1,000,000 tons of material.

There is no satisfactory explanation of the ruined walls *inside* the ellipse. There are circles, smaller ellipses, a lower wall running parallel to the outside one and a tower, 10 m high on a base with a diameter of 6 m, in the right-hand corner (of course an ellipse has no corners, strictly speaking). I could make no sense out of the tower. It has no entrance, no steps or windows, and its exterior wall is completely packed with stones on the inside.

The English archaeologist, Gertrude Caton-Thompson, who was in charge of excavations in 1929, thought there was a tomb under the tower. Digging took place, but no tomb was found. So the tower stands there among the other buildings, even though it has no apparent purpose.

A less spectacular site, called the 'valley ruins', extends around the ellipse. However, I found no signs of a valley. The ruins are scattered over the same plain on which the large ellipse lies. And as is only appropriate here, luxuriant colourful plant life flourishes between the stones.

The large ellipse and the valley ruins are dominated by a third complex, which lies on a hill and is called the 'Acropolis'. The natural features of the ground have been used with extraordinary ingenuity. Walls have been built wherever there are gaps in the rock. The thickest, outer walls are 7.50 m high and 6.70 m wide at the base, and although they taper upwards they are still 4 m wide at the top! The construction workers must have had good heads

This conical tower which stands in the right-hand 'corner' of the ellipse has no openings at all. No entrance. No exit. No windows.

for heights, because some sections of the walls are built on sheer rock faces. *These sections* of the Acropolis must have been easy to defend—if Zimbabwe really was a fortress.

Digs on the hill have revealed small gold bracelets, glass beads and eight birds made of soapstone, the mineral saponite, which feels like soap in its dry state. These 'Zimbabwe birds' add to the mystery of the site. 30 cm high, they were probably perched on columns originally.

There are geometrical patterns on the floor of the Acropolis. Looking down on the valley ruins and the great wall, the view is breathtaking.

Some of the stone blocks, which are over 15 m high, seem to have been worked by human hands, others could have been dressed mechanically. I know monoliths in Peru which bear similar traces. There, above the Inca fortress of Sacsayhuaman, it looks very much as it does here in

The 'Acropolis' towers over the cliff face with its clefts and chasms.

Zimbabwe—as if giants had once been toying with the blocks of stone. Today a narrow zigzag stairway leads up to the monoliths. Anyone ascending in the noonday heat without a guide should beware of snakes.

Cecil Rhodes (1853–1920), the founder of Rhodesia, visited Zimbabwe and was extremely interested in the numerous theories about the origin of the buildings. He opted for the biblical version, according to which Zimbabwe was Ophir, the land of gold.

About the same time, the archaeologist J. P. Went supported the view that Zimbabwe had been built by the Arabs. Today this view is still shared by R. Gayre, who says that Bantu negroes have never built monolithic buildings anywhere else, so why here? Gayre uses the gold trade to justify his hypothesis. The Arabs mined here in pre-Islamic times and built Zimbabwe to protect their treasures. As

One of eight Zimbabwe birds which ask: where do we come from?

regards the elliptical wall, he points to a similar seventeenth-century wall, which is located in the Yemen.

According to some estimates, as much as 600,000 tons of gold a year were excavated in Zimbabwe's heyday. Today Rhodesia's annual gold production is a mere 16 tons.

Everything is problematical; uncertainty reigns. Zimbabwe really does seem to have no history.

As I like to imagine that our ancestors, whether black or white, thought about practical matters very much as we do, I never felt at all happy about the dreamland of Ophir idea, when actually faced with the massive buildings. Why not?

If a garrison is supposed to have been there to guard the transport of gold, the soldiers would certainly have lived in the formidable Acropolis. They could have overlooked the plain from up there.

On the other hand the ellipse of the big ruins in the plain makes no sense. There was no commanding all-round view, nor any of the features which defenders in all ages have needed: towers, battlements, embrasures! It was not even possible to climb the walls of the ellipse; there were no steps leading to the top or projections from the wall to scramble up. As a fort the great ellipse is a washout.

Why, for heaven's sake, did African negroes drag hundreds of tons of granite here and then break them up to build this monumental edifice?

I could not get this question out of my mind; it followed me daily on my visits to the various ruins . . . until a map of the complex on the wall of the Zimbabwe Museum gave me an idea!

Inside the great ellipse, the massive conical tower in the 'right-hand corner' has a significant position. Ellipse and tower—surely they resemble, with slight distortions, the Sirius model which was discovered among the Dogon negroes in the West African Republic of Mali?

The scholar Robert K. G. Temple has clearly proved

Walls. Walls. Walls. But no doors, no battlements, no embrasures, no steps. What was the purpose of this massive edifice?

that the Dogon tribe has known about the Sirius system down to the last detail from time immemorial.

Sirius A is the main star in the constellation of Canis Major. A tiny invisible neutron star, Sirius B, revolves around it in an elliptical orbit. This orbit around Sirius A is clearly recognisable in the 'bottom right' of the Dogons' stone drawings.

The Dogon assert that they got their enormous astronomical knowledge from a god called Nommo. But not only did Nommo tell the negroes about the orbit of the *invisible* Sirius B around Sirius A, he also supplied the names and orbital data of some other planets in the Sirius system. For example, there is a 'shoemaker planet' and a 'planet of the women'—knowledge which modern astronomy does not yet possess. *It* only knows that Sirius B revolves round Sirius A in an elliptical orbit lasting fifty years.

Standing in front of the plan in the Zimbabwe Museum, I felt strongly that there was a visual parallel. Surely the great ellipse of Zimbabwe with the conical tower at the 'bottom right' resembles the traditional Dogon Sirius model? Do the inexplicable ruined walls *inside* the great ellipse trace the orbits of the 'shoemaker planet' and the 'planet of the women'? Why else does an apparently meaningless wall run parallel to the elliptical great wall for a full third of its length? It cannot have served a defensive purpose, any more than the other round or spiral-shaped walls inside the ellipse.

You only need to fly at a moderate altitude to see that the great ellipse of Zimbabwe with the massive tower in the bottom right-hand corner is almost identical with the Dogon Sirius model.

The question is whether there is an ideological connection between the Zimbabwe complex and the Dogon Sirius model, in addition to their astonishing visual similarity?

In all places at all times religion has been the driving force that spurred men on to superhuman achievements. Signs for the gods all over the world were of religious origin. Religious impulses inspired megalithic temples and the pyramids, not to mention Arabic mosques and Christian cathedrals. Incas and Mayas built their step pyramids and temples in honour of the gods. In all the leading religions in

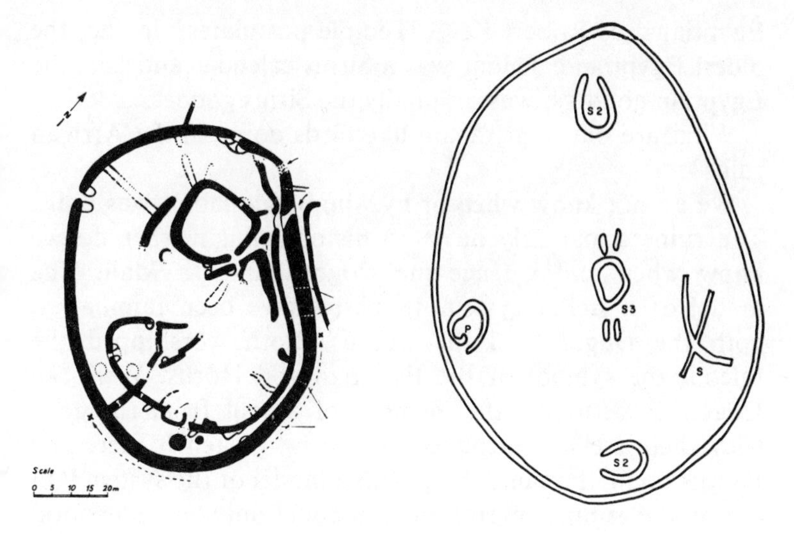

Diagram of the ruins of Zimbabwe compared with model of Sirius made by the Dogon negroes from Mali.

the world even the poorest of the poor collected gold and precious stones to ornament the symbols of the gods. Heathen or Christian, it makes no difference; man has always built and made sacrifices in order to honour a god.

We must ask ourselves whether the Zimbabwe blacks, worshippers of their god Nommo, built a gigantic edifice in memory of his origin in his home amid the stars—a model of the Sirius system. Did religious fervour drive them to undertake the monumental work in order to express in stone their hope that Nommo would return? Did they want to signal to their god, 'Here we live and wait for you!'

The eight Zimbabwe birds from the Acropolis are very similar to the sacred falcons of the Egyptian god Horus, who was originally a celestial god. He was symbolised by a falcon with outspread wings.

Did the Dogon acquire their ancient wisdom from the

Egyptians, as Robert K. G. Temple postulates? In fact the oldest Egyptian calendar was a Sirius calendar and Isis, the Egyptian goddess, was originally the Sirius goddess.

What are the eight falcon-like birds doing in the African ruins?

We do not know when or by whom Zimbabwe was built. The ruins apparently have no history. But neither do we know when and whence the Dogon came to Mali. The model of a stellar system seems to have been familiar to both the Dogon and the Bantus. Both worshipped the falcon, the symbol of the Egyptian god Horus. Were the Dogon perpetuating the memory of a visit from the gods with their stellar legend of the Sirius system? Were the Bantus doing the same thing with a model of the system laid out on the ground, even though it could only be understood from above?

I do not claim that my contribution provides the solution to the Zimbabwe mystery. I only know that so far no one has said anything at all satisfactory. As the great ellipse of Zimbabwe was not a fortress—the fortress is 100 metres higher up on the Acropolis—it must have been either a residence or a kind of temple. But the residence theory falls down because no signs of occupation for this purpose were found; there are no kings' names or decorations on the rather barbaric masonry. Nothing resembling a throne was found. No rooms to indicate that men once lived here. And what could a king have had to do with the conical tower inside the ellipse or the pointless second wall running parallel to it?

If the fortress and residence theories are eliminated, we are left with the assumption of a religious cult. When I was staying in Zimbabwe, I could easily imagine a procession of chanting Bantu negroes advancing through the passage between the parallel walls towards the conical tower and worshipping the god Nommo from the Sirius system.

The solutions to the problem of Zimbabwe that have been put forward so far are no more than speculations. That is why I add my own. It is just as valid as any other speculation about Zimbabwe.

After all my studies and travels, I keep on thinking about the grey-haired gentleman from Athens. I must beg his forgiveness.

I could easily imagine a procession of chanting Bantu negroes winding their way through the passage between the walls . . .

Communiqué

The Rongomai legend of the New Zealand Maoris says:

'There was a war between the ancestors of the Nga-Ti-Hau and another tribe. The bad tribe dug in in a Pa (fortified village). The priests of the Nga-Ti-Hau tribe begged their god Rongomai for help, because the bad tribe had stolen a sacred relic. At noon the god Rongomai came through the air. He was like unto a twinkling star, or a comet or a flame of fire. He flew until he was directly above the Pa and then dived swiftly down on the Maray (village square). The earth was whirled up in heaps and scattered, and the noise was that of thunder. The Nga-Ti-Hau warriors hailed their god Rongomai with shouts of joy and occupied the Pa at once.'

Source: John White: *Ancient History of the Maori*, New Zealand, 1887.

6: Right Royal King Lists

Large reference books such as lexicons and encyclopedias have the invaluable advantage for the average reader of supplying accepted knowledge succinctly and neatly, without 'ifs' and 'buts'.

Testing this under the heading 'Sumerians', I found data like these in various learned tomes:

> 'Sumerians, the inhabitants of the region (Mesopotamia) between present-day Baghdad and the Persian Gulf. A people of unknown race, whose existence since the beginning of the third millennium B.C. can be proved on linguistic grounds. When and whence the Sumerians migrated to Babylon, in the region between the Euphrates and the Tigris, *has not yet been discovered*.'

Or:

> 'Sumerians, the inhabitants of central and southern Mesopotamia from the fourth to the second millennium B.C.'

Or:

> 'The origin of the Sumerian people is *uncertain*. They may have come from the mountains of the east or from the sea. The only *definite* fact is that they were settled in Mesopotamia at the beginning of history.'

It is unanimously admitted that we do not yet know where this people came from. Their traces might well have been blown away by the winds of time, but for the native intelligence of this people with the Nansen passport*. They invented 42 alphabetical cuneiform characters and so turned ephemeral speech into permanent writing.

*Travel document for stateless or quasi-stateless refugees.

Excavations south of Baghdad yielded more than 30,000 clay tablets dating back to the Sumerian epoch.

It is extraordinary, even uncanny, but only 100 years ago not even the name of Sumer was known! The Assyriologist Jules Oppert (b. Hamburg 1825, d. Paris 1905) was the first to locate the land of Sumer by deciphering the cuneiform script. That was in 1869.

Professor Samuel Noah Kramer, Assyriologist at the University of Pennsylvania, found out from drawings on clay tablets that the cart-wheel and the sailing ship were among the Sumerians' technical achievements, that they were governed by well-organised authorities and that we still make use of their astronomical knowledge today—where on earth did they get it? They knew everything from the sixty-second minute to the solar year.

Just as Europe was leaving the Neolithic Age, the Sumerians had already thought of stamping documents, bills, etc., with the impression of a seal of office. They invented the cylinder seal. These seals, only two to six centimetres long, were worn on necklaces so that they were always to hand. Tax collectors used them to give receipts. Seals, some of them of great beauty, have now been used globally for 4,000 years. O those Sumerians!

What further information can we add to the unknown people's file?

Colour of hair: very dark. Inscriptions mention 'black heads'.

Race: before the Sumerians emerged between the Euphrates and the Tigris, Semitic tribes lived there, but they themselves were definitely not Semites, nor were they of negroid descent.

From depictions on Sumerian reliefs we can recognise an Indoeuropean mixture, which obviously spread far and wide, for Sir Arthur Keith noted:

> 'One can still see the facial traits of the Sumerians in the East, among the inhabitants of Afghanistan and

Baluchistan, and as far as the Indus valley, about 2,400 km away.' (43)

Wherever they came from, the Sumerians obviously brought with them—as modern scholars know—a polished culture and a complete civilisation which was so infinitely superior to the native tribes that they could not stand up to it and were wiped out.

They were also conscious of their superiority, for they described themselves in numerous creation myths as 'the authentic founders of civilisation', as men born to serve their divine creators. (44) 'With the help of their gods, especially Enlil, the "King of Heaven and Earth", the Sumerians transformed a flat, arid, windswept land into a blossoming, fertile kingdom' (Kramer).

How was it possible for a highly developed culture to emerge from nothing at least 4000 years BC?

Who taught them urban architecture (in a crash course?!)?

Who taught them how to organise their twelve city states so efficiently?

Where did they get the engineering knowledge to canalise their country? For that is how they protected the harvests from the raging floods which constantly overflowed the banks of the Euphrates.

Where did they get the mathematical ability—confirmed on cuneiform tablets—to handle squares, cubes, reciprocal values, roots, powers and even some abstruse Pythagorean calculations? How could they calculate areas and circles? Who told them that a circle is divided into 360 degrees? Who gave them that unit of measurement?

Today such evidence of Sumerian history can be visited at archaeological sites in Mesopotamia or in the large collections in the British Museum and the Louvre. They take your breath away. In my opinion the Sumerians should have left traces of their technology, culture and religion on what was presumably a lengthy journey to Mesopotamia. They did not, otherwise we should know where they came from.

Some archaeologists think that the Sumerians did not immigrate, but developed in their native land between Baghdad and the Persian Gulf. Excavations at Uruk actually yielded documents with strange lists giving the names of things and concepts such as house, bird, fire, temple, god, heaven, rain, etc., etc., as if a teacher was instructing primitive peoples and saying: 'Take a good look. That is the name of this thing!' Aids to evolution?

Strictly speaking the confusion about the Sumerians could be settled. While the first dynasty ruled in Isin, the ancient royal city south of Babylon, from 1953–1730 BC, a chronology of the past, the 'King Lists', was drawn up. Copies have been preserved. In the fourth or third century B.C. the Babylonian priest Berossus transcribed them, somewhat imaginatively, into Greek. Nevertheless, they offer a serviceable staircase back into time.

But in 1932 Sumerologists were astounded, confused and delighted. The *original* King Lists were found at Khorsabad, the Iraqi town near Mosul in the valley of the Tigris. Now scholars had authentic names and dates to deal with.

The oldest and most precise dynastic list is called *Ancient Babylonian King List WB 444* in archaeological literature. It is on a block 20.5 cm high—big enough to trace back the series of mysterious original kings to the distant time of the creation of man.

The continuation is known as *Babylonian King List A*. The beginning, the names and dates of the first dynasty, is illegible. *King List B* fills the gap. It contains the names of the kings of the first Babylonian dynasty (1830–1530). Thus we know, insofar as they are legible—the teeth of time have been at work—the names of the Sumerian and Babylonian rulers and the dates of their reigns!

Was the Sumerian mystery solved by the lucky find of the King Lists? Not a bit of it! That was when the trouble really started.

According to WB 444, the first ten kings ruled from the creation to the Deluge, a total of 456,000 years. Yes, you'd better read that twice, it's not a misprint! In words, four hundred and fifty-six thousand years! After the Deluge 'the kingship came down from heaven again'. The 23 kings who then succeeded each other to the throne still managed to clock up reigns lasting 24,510 years, three months and three and a half days. Quite a few years!

Although officially the King Lists are 'catalogues of the kings and their reigns arranged by dynasties,' (45) scholars felt that there was something wrong somewhere. Only Sir Leonard Woolley who went on busily excavating in Sumerian soil came to believe in the King Lists, even though he himself could not explain them (43). Our archaeologists find the years given for the various reigns *too* astronomical.

I admit that according to old standards they do not admit of any firm conclusions.

But before I put forward my speculations about these lengthy reigns and their problems, I must at least introduce my readers to a selection from the King Lists. The list going back to the creation of man would fill several pages and be completely unnecessary in this context.

Examples from King List WB 444:

When the kingship came down
from heaven,
the kingship was in Eridu.
Alulim was king in Eridu.
He reigned *28,800* years.
Alalgar reigned *36,000* years.
Two kings,
They ruled *64,800* years.
In Bad-Bad-tibira En-men-lu-anna
ruled *43,200* years,
En-men-gal-anna
ruled *28,800* years.
En-men-dur-anna was king
in Sippar, he ruled *21,000* years.
One king,
he ruled his *21,000* years.
In Suruppak Ubar-tutu was king,
he ruled *18,600* years.
Five cities,
Eight kings,
they ruled *241,200* years.
The Deluge came down.
After the Deluge came down,
and the kingship came down
from heaven (again),

God *Dumuzi*, the shepherd,
ruled *36,000* years.
Three kings,
they ruled their *108,000* years.
In Larak En-zib-zi-anna
ruled *28,000* years.
One king,
he ruled his *28,800* years.
Atab ruled 600 years,
Atab's son ruled
840 years.
ETANA, the shepherd,
who rose up to
heaven,
who fortified the countries,
was king.
He ruled *1,560* years
Balih,
son of Etana,
ruled *400* years.
Tizkar, son of Samug,
ruled *305* years.
Ilta-sadum
ruled *1,200* years.
(Mes)-kiag-ga(ser),
son of the Sun God,
was high priest
the kingship was in Kiš.
In Kiš Ga-ur was king,
he ruled *1,200* years.
Gulla-Nidaba-anna-pad
ruled *960* years.
Zukakip
ruled *900* years.
(and king), (he ruled)
324 years.
The divine Lugal-banda,
the shepherd,
ruled *1,200* years.
God Dumu-zi, the fisherman,
his town is Kua,
ruled *100* years.
The divine GILGAMESH,
his father was a Lillu demon,
high priest of Kullab,
ruled *126* years.
Ur-nungal, son of
GILGAMESH,
ruled *30* years.
Utul-kalamma, son of Ur-nungal,
ruled *15* years,
Labaser
ruled *9* years.

The King Lists are also lists of gods, to some extent. Kings, who were not only worshipped as gods by the Sumerians, but also recognised as teachers, figure in them. Gilgamesh, Etana and Enkidu were heroes of famous epics which bear their names. Names from the King Lists were also found on cuneiform clay tablets and seals, which proves that the lists were not a product of the imagination of one or more chroniclers. The kings did exist and their influence was 'stamped', as it were, on everyday life.

But what is the meaning of the incredibly long reigns of the crowned heads?

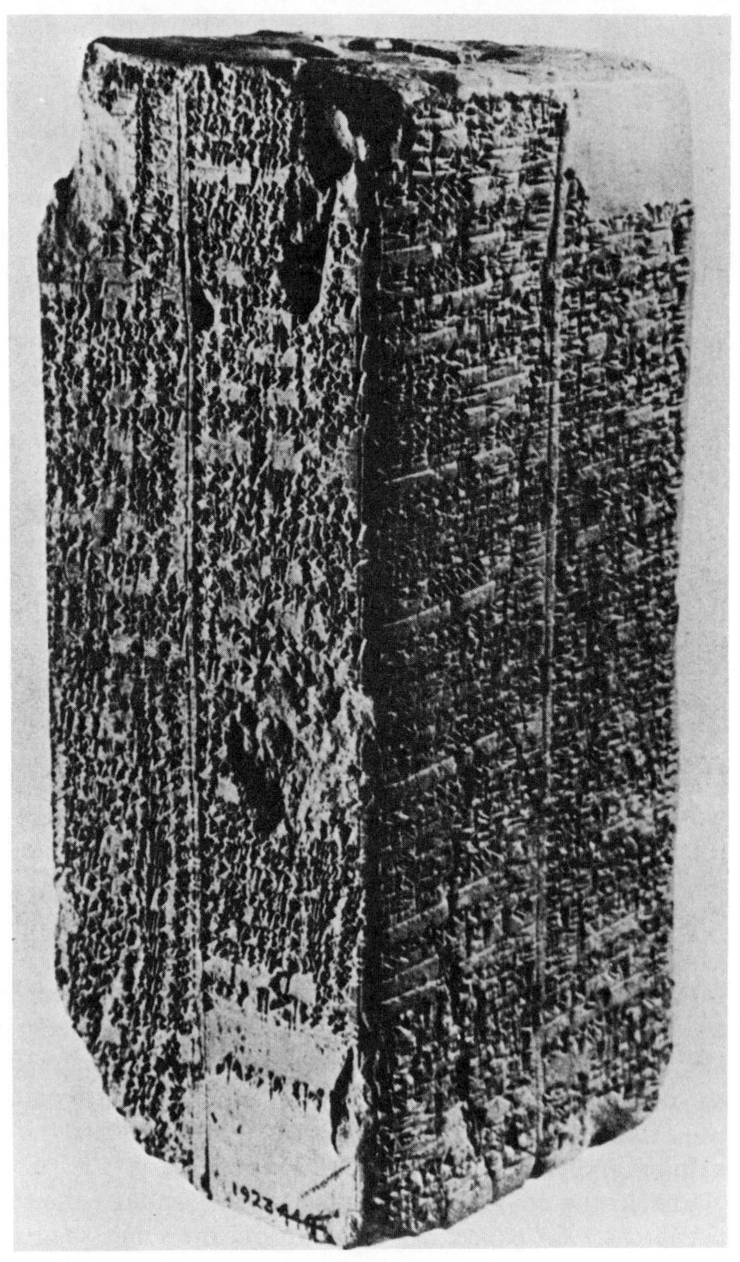

The ancient Babylonian King List WB 444.

Friedrich Schmidtke (46) writes of the confusion in which Sumerologists find themselves:

> 'At first glance it looks as if the dynasties ruled one after another, which would lead to impossible consequences as far as the duration of Sumerian history is concerned.'

In scholarly circles, they wonder what can have impelled the chroniclers of the King Lists to put down 'such impossible figures' (46). Before Professor Schmidtke presents the tables with the names and dates of the Sumerian and Babylonian kings, his sense of resignation shows through:

> 'The contents of WB *before this* lie in the realm of saga and need not concern us here, however interesting the antediluvian dynasties are for the history of religion.'

Should these startling datings be relegated to the kingdom of fable and legend? Should we make things so easy that anything that is not immediately explicable goes to the shunting yard and is then addressed to the terminus labelled fable + legend?

Does this mean that everything that we do not understand must be attributed to the great magician, chance?

WB 444 registers *ten original* kings from the creation of the earth to the Deluge. Altogether they reigned for 456,000 years.

The Bible names *ten patriarchs* from the creation of Adam to the Flood (45) and these gentlemen, too, lived to an astonishingly old age.

Pablo Picasso, who had his daughter Paloma when he was 68, was a young man compared with Adam, the first man, who is supposed to have produced his first son at the age of 130. And in comparison with Adam, who is reputed to have lived for more than 900 years, Picasso was a mere strippling when he died aged 92.

Enoch, the antediluvian prophet and seventh of the ten patriarchs, can expect an even more astonishing span of years. He did not die at all, but was taken to heaven by

God. Methuselah, his son, the patriarch immediately before the Flood, died at the blessed age of 969.

If you ask gerontologists about the reliability of reports of abnormally high ages, American and Russian specialists are unanimous in saying that nature has fixed the age of man at 110 to 120 years. The constantly publicised reports of the legendary Bulgarian shepherd who has reached the age of 150—once again—belong to the realm of fairytales. When anyone tries to justify the truth of such reports, there are no supporting documents of any kind that show the date of birth. Old wives' tales . . .

Our life span is determined by the functioning of the 15 milliard cells in our bodies. The cells divide in the course of life and keep on rebuilding our bodies. With each division we slowly approach our end—from the age of 20 onwards, for the intensity of cell renewal is over after 30, or at the most 50 divisions.

Man's expectation of a 'biblical' age of 110 to 120 years is still only a wish, a dream . . . until gerontological research succeeds in slowing down cell degeneration. Scientists who have examined mummified tissue from early times say that men subject to different physical laws have *never* existed.

Knowing these undisputed facts, I cannot help asking why the Sumerian and biblical chroniclers 'invented' such astronomical ages for their forefathers—ages that they could not have reached according to modern science.

In the hill of El Obeid near Ur of the Chaldees, Sir Charles Leonard Woolley found a limestone tablet with these words:

> 'Dedicated by A-anni-tadda, King of Ur, son of Mes-anni-padda, King of Ur (43).'

This Mes-anni-padda appears in the King Lists as founder of the third dynasty after the Deluge.

It is astonishing that certain names appear in the King Lists at different times in different dynasties, as if they had ruled more than once and had only vanished for a few centuries or millennia in the interim.

Here is an example. The Babylonian king Nabu-na'id (555–538 BC) says on a tablet found in the Temple of the Sun at Sippar:

> 'For the sun god, the judge of heaven and earth, did I rebuild the Temple of the Sun, his house at Sippar, which Nebuchadedsae, a former king, built, and whose ancient foundations he sought, but did not find. With the passage of 45 years, the walls of that house had collapsed; at this I was sore afraid, I fell to my knees, terror seized me and my face was disturbed.
>
> While I removed the god's image from inside the temple and put it in another temple, I demolished that house, I sought the ancient foundations and I made the groundlevel 18 ells deeper, and the sun god, the great lord of the Temple of the Sun, let me see the foundations of Naram-sin, the son of Sargon, which no former king had seen for 3,200 years . . .*
>
> On the foundations of Naram-sin, the son of Sargon, laid I the building stones in points going neither in nor out.'

King Nabu-na'id clearly states that the foundations he sought so zealously and then found 18 ells below groundlevel showed that his royal ancestor Naram-sin first built the Temple of the Sun 3,200 years *before his time* (i.e. about 3,800 BC).

A remarkable thing is that *the same Naram-sin*, like his father Sargon, reappears in the King Lists at quite different times.

One more example: according to King Lists A and B *Hammurabi* reigned some 700 years before Burnaburias I. Sumerologists say that this is out of the question.

> 'The statement that Hammurabi lived 700 years before Burnaburias I is quite impossible.' (46)

*The dots indicate omissions where the original is illegible.

Why is it impossible? That is exactly what it says in the laboriously compiled King Lists!

The names of various kings are perpetuated on cuneiform clay tablets found on many sites. They state irrevocably that these monarchs did in fact rule. The inscriptions with the names are documentary evidence. The King Lists tell us which rulers were at the helm and for how long they reigned.

Now the real jigsaw puzzle begins!

Sumerologists are seriously trying to ascertain the exact chronological sequence of the dynasties and their kings. They think they can fix the dates of *one* king's reign on the basis of some inscription found at some site or other.

One date is recorded and a chronological sequence running backwards and forwards is built up from it. King X was succeeded by King Y. King Y was killed in a war by King Z. Consequently King X must have lived *before* King Z.

And then this distant period of history plays the zealous scholars a dirty trick! Kings X, Y and Z suddenly appear on quite different clay tablets in a different series of successions to the throne and in a quite different context.

What do they do so as not to spoil the painstakingly compiled family trees?

They do what archaeologists usually do in such cases. They blame the old chroniclers for everything. Those gentlemen could not count, they say. They wrote the kings and their dates down next to each other, instead of one after another, they say. They would have us believe that the chroniclers were pretty stupid in general.

But after these limp evasions, the fact remains that the defenceless chroniclers accurately noted one reign after another in the original King Lists! It is incomprehensible why such unreliable historians were also at work in the Old Testament—listing the ten antediluvian patriarchs.

At first sight I do not find the assumption that the *same* chroniclers might have been at work too far fetched. The

Bible and historical research agree that the young Moses, later liberator of Israel and founder of the religion of Yahweh, grew up and was educated at the court of a Pharaoh. He almost certainly had access to the splendid libraries of the second millennium BC.

Did Moses have a look at the Sumerian King Lists? Did he commit the information to his phenomenally good memory and hand it on as part of the oral tradition? If so, why did he not take over the *same* figures for the ten Old Testament patriarchs that the Sumerians had incribed next to the names of their ten antediluvian kings? One can toy with the idea that the same source may have been tapped for both our Sumerian and biblical ancestors, but on careful examination there is only one common feature: the incredible ages attributed to both kings and patriarchs. That is not enough to explain the phenomenon.

I should like to introduce three speculative explanations into the discussion:

1 From time to time the antediluvian kings were invited by extraterrestrials on journeys to other solar systems.

Impressive descriptions of such journeys are preserved in both the Zohar, the main work of the traditional Jewish Cabbala, and in the Book of Enoch, which the early Abyssinian church accepted as canonical.

The technical historian Professor Richard Hennig described certain parts of the Sumerian legend of Etana as 'the oldest flying story in the world'. Written in cuneiform script between 3000–2500 BC, this aerial journey was also depicted pictorially on cylinder seals. The Sumerian Epic of Gilgamesh describes the hero's dreamlike journey to the dwelling-place of the gods.

Aerial journeys to distant worlds are not exclusive fairy-tales confined to the peoples of the Middle East. They are also found in the Indian national epic *Mahabharata* and in the *Ramayana* written between the fourth and third centuries BC. They occur in Nordic myths and Red Indian traditions. There is no national copyright for heavenly journeys with the 'gods'.

Since Albert Einstein (1879–1955) developed his special theory of relativity, enormous figures for the beginning and end of 'life' have become explicable. After physical experiments, Einstein's theory had proved to be a fact of natural law.

The eternal natural law of time dilation says neither more nor less than that for astronauts on an interstellar spaceship travelling close to the speed of light time passes *more slowly* than it does for the observers who stay behind at the launching ramp.

Since Einstein time is no longer a fixed dimension; it can be manipulated by energy = speed.

How do the incomprehensible Sumerian dates in the King Lists look in the light of this knowledge?

The surviving Sumerian inscriptions do not tell us about vague matters of foreign policy which could have got muddled up with the net of dates. They inform us drily about concrete events like the construction of palaces or temples, which were obviously erected for the 'gods' dwelling in their midst. This practice is not at all surprising, for the Sumerian kings looked on themselves merely as representatives of the 'real' gods. These 'gods' installed the kings personally, and they followed the same procedure again after the worldwide Deluge. When the floods had ebbed away, when our blue planet had become uninhabitable, 'the kingship came down from heaven again'. That is what it says in the King List.

If this is accepted as fact, it is no longer so absurd to presume that the god-kings were either living extraterrestrials or had at least been taken backwards and forwards on flights to other systems by extraterrestrials.

The 'impossible' figures for the royal reigns and the assurance that the kingship came down from heaven entitle us to suspect that we are not dealing with things of this world.

When we know the effect of time dilation, the total of 456,000 years for which the ten kings ruled is no longer so disturbing. It's a bagatelle!

2 The extraterrestrials (gods) produced sons and daughters after mating with the children of earth. So the prophet Enoch claims. That is what it says in the Lamech Scroll, which is over 2000 years old and was found in 1947 in the settlement of Chirbet Qumran near the Dead Sea.—The Sumerian god Enlil, who reigned in Nippur, seduced the delightful Ninlil and got her with child.—Even Genesis mentions marriages between the 'sons of God' and the 'daughters of men'.

The products of this unusual act of procreation could undoubtedly have told amazing details about their split inner lives on the psychiatrist's couch. These hybrids emulated their 'divine' creators, probably out of envy, because according to all the traditions the 'gods' were immortal, whereas they had to die like the rest of the inhabitants of earth.

The descendants of the 'gods' produced on our earth became mortal, because once the extraterrestrials departed definitively, they no longer had a chance to take part in interstellar spaceflights at high speeds. They could no longer avoid the process of ageing.

It is quite understandable that the sons of the gods strove to overcome this vulgar mortality; understandable, too, that they ruled as long as humanly possible and insisted on their royal prerogatives. Even then the ruling classes enjoyed their power.

What was the Sumerian paradise Dilmun, that divine garden 'in which there was neither sickness nor death'?

What was the 'herb of immortality', which Utnapishtim, an ancestor of the hero Gilgamesh, knew. An immortal himself, he lived on an island 'on the far side of the sea of the dead'. What was this 'plant of eternal youth'?

Utnapishtim, a survivor of the Deluge, entrusted its secret

to Gilgamesh. He told him that immortality was a property of a plant from the fresh water sea. Gilgamesh obtained the plant and was going to give it to his nearest relatives to eat. On his way home, he stepped into a spring to wash himself. Then a snake came and stole the precious herb. Gilgamesh wept.

Did the sons of the gods and/or the antediluvian kings know medicaments which could drastically slow down the degeneration of the cells? Preparations which preserved the vital functions longer?

So far the herb of immortality has not been found. Geriatric research is still looking for it. For us.

3 Did the sons of the gods and/or the antediluvian kings have themselves mummified and laid in tissue-preserving containers guarded by priests who woke them up again after centuries had gone by?

Did they know methods of refrigeration at low temperatures which, unlike all known experiments in this direction, excluded crystallisation of the cell walls and cell nuclei? Is this the reason for the constantly recurring claims that the 'gods' were always 'present' in the temple?

The high-ranking priests knew that the gods lived among them physically at times, that they were the real owners of the cities and only left every-day administration to the kings they had installed (47). The priests feared both the return of the extraterrestrial gods and the reawakening of the sleeping sons of the gods.

The temples were originally intended as places for *actual* encounters with *genuine*, living gods. It was only much later, when the gods no longer returned and the sleeping sons of the gods did not answer reveille, that the priests used all kinds of tricks to keep king and people docile. Statues were set up in the temples as representatives of the heavenly ones.

Perhaps these three speculations may help to solve the mystery surrounding King List WB 444. The facts stored in the data bank are too precise to be overlooked.

Communiqué

On September 5, 1978, Dr Knut Oppenländer of Ludwigshafen am Rhein drew my attention to a curiosity he had come across in a book belonging to one of his children, *Tatsachen/Die verblüffendsten Rekorde der Welt*. If my correspondent's item was a world record, it would have to be in the *Guinness Book of World Records*. I went to my library, took out the 1978 edition and found, on page 207, the longest name in the world.

The text was identical with the quotation from the German book sent by my correspondent. So here is the longest name:

Adolph Blaine Charles David Earl Frederick Gerald Hubert Irvin John Kenneth Lloyd Martin Nero Oliver Paul Quincy Randolph Sherman Thomas Uncas Victor William Xerxes Yancy Zeus Wolfeschlegelsteinhausenbergerdorffvoralternwarengewissen-
shaftsschaferswessenscahfewarenwohlgepflegteundsorgfaltig-
keitbeschutzenvonangreifendendurchihrraubgierigfeindewelche-
voralternzwölftausendjahresvorandieersheinendenvanderersteer-
demenschderraumschiffgebrauchlichtalsseinursprungvonkraft-
gestartseinlangefahrthinzwischensternartigraumaufdersuchen-
achdiesternewelchegehabtbewohnbarplanetenkreisdrehensichun-
dwohinderneurassevonverstandigmenschlichkeitkonntefortpflan-
zenundsicherfreuenanlebenslanglichfreudeundruhemitnichtein-
furchtvorangreifenvoneinanderintelligentgeschöpfsvonhinzwisch-
ensternartigraum, Senior. who was born at Bergedorf, near Hamburg on February 29, 1904.

A leg pull? Not a bit of it. That is the name in the passport of the man who was born near Hamburg on February 29, 1904, and later emigrated to America. As you will understand, the full name was very inconvenient for visiting cards and notepaper. Until recently the gentleman used only his second and eighth Christian names and the first thirty-five letters of his surname. Today—he lives in Philadelphia, USA—he has shortened his name to Wolfe + 585, Senior. That is something he can live with.

The editors of the *Guinness Book of World Records* have included this extraordinary name without realising that it is a communication in mediaeval German. This is how it reads when modernised:

> 'A long time ago there lived conscientious shepherds who tended their sheep carefully. Then rapacious enemies appeared

before the first earth men. This was 12,000 years ago. The spaceships used light as their source of energy. In search of habitable planets they had made a lengthy journey in stellar space. The new race propagated itself with intelligent mankind. They rejoiced in their life, without fear of attacks by other intelligent creatures from space.'

A mediaeval ancestor of Mr Wolfe + 585, Senior, must have had knowledge about our remote human past which he wanted to hand down to future generations by including it in the endless surname. So that some day someone would stumble over this tapeworm of letters.

In spite of the abbreviation to a 'telegraphic' name, Mr Wolfe has remembered something of his ancestor's wish; with his '+ 585'; he struck out exactly 585 letters.

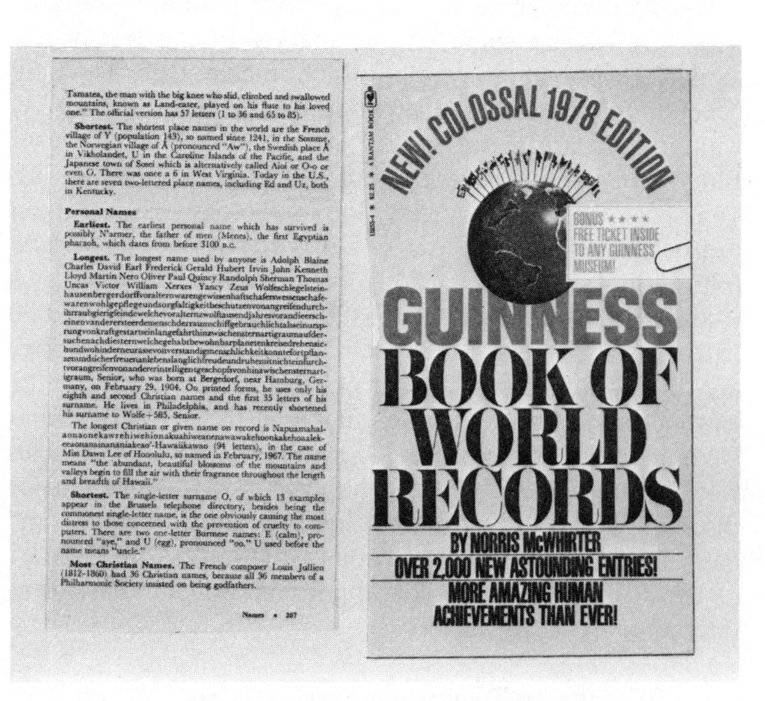

Tamatea, the man with the big knee who slid, climbed and swallowed mountains, known as Land-eater, played on his flute to his loved one." The official version has 57 letters (1 to 36 and 65 to 85).

Shortest. The shortest place names in the world are the French village of Y (population 143), so named since 1241, in the Somme, the Norwegian village of Å (pronounced "Aw"), the Swedish place Å in Vikholandet, U in the Caroline Islands of the Pacific, and the Japanese town of Sosei which is alternatively called Aioi or O-o or even O. There was once a 6 in West Virginia. Today in the U.S., there are seven two-lettered place names, including Ed and Uz, both in Kentucky.

Personal Names

Earliest. The earliest personal name which has survived is possibly N'armer, the father of men (Menes), the first Egyptian pharaoh, which dates from before 3100 B.C.

Longest. The longest name used by anyone is Adolph Blaine Charles David Earl Frederick Gerald Hubert Irvin John Kenneth Lloyd Martin Nero Oliver Paul Quincy Randolph Sherman Thomas Uncas Victor William Xerxes Yancy Zeus Wolfeschlegelstein-
hausenbergerdorffvoralternwarengewissenhaftschaferswessenschafe-
warenwohlgepflegeundsorgfaltigkeitbeschutzenvonangreifendurch-
ihrraubgierigfeindewelchevoralternzwolftausendjahresvorandieersch-
einenvandererersteerdemenschderraumschiffgebrauchlichtalseinursp-
rungvonkraftgestartseinlangefahrthinzwischensternartigraumaufder-
suchenachdiesternwelchegehabtbewohnbarplanetenkreisedrehensic-
hundwohinderneurassevonverstandigmenschlichkeitkonntefortpflan-
zenundsicherfreuenanlebenslanglichfreudeundruhemitnichteinfurch-
tvorangreifenvonandererintelligentgeschopfsvonhinzwischensternart-
igraum, Senior, who was born at Bergedorf, near Hamburg, Germany, on February 29, 1904. On printed forms, he uses only his eighth and second Christian names and the first 35 letters of his surname. He lives in Philadelphia, and has recently shortened his surname to Wolfe + 585, Senior.

The longest Christian or given name on record is Napuamahal-
aonaonekawrehiwehionakuahiweanenawawakehoonkakehoaalek-
eeaonanainananiakeao'-Hawaiikawao (94 letters), in the case of Miss Dawn Lee of Honolulu, so named in February, 1967. The name means "the abundant, beautiful blossoms of the mountains and valleys begin to fill the air with their fragrance throughout the length and breadth of Hawaii."

Shortest. The single-letter surname O, of which 13 examples appear in the Brussels telephone directory, besides being the commonest single-letter name, is the one obviously causing the most distress to those concerned with the prevention of cruelty to computers. There are two one-letter Burmese names: E (calm), pronounced "aye," and U (egg), pronounced "oo." U used before the name means "uncle."

Most Christian Names. The French composer Louis Jullien (1812–1860) had 36 Christian names, because all 36 members of a Philharmonic Society insisted on being godfathers.

Names • 207

Sources: 1) *Guinness Book of World Records*, 1978.
2) *Tatsachen/Die verblüffendsten Rekorde der Welt*, pp. 157 *et seq.*, Munich-Vienna, 1976.

7: Prophet of the Past

Ten years ago my firstling *Chariots of the Gods* topped the best-seller list in nearly every country in the world. After the initial stupefaction and obligatory contempt, there was a global storm of enthusiasm (and indignation).

What an incredible amount has been written about 'gods from outer space' since 1968!

During these ten years, 321 (!) books dealing favourably with 'my' theme came on the market in the free world alone. They include works which tackle the theory in general, others which concentrate on the specific country, while others are about special aspects, for example Josef Blumrich's *The Spaceships of Ezekiel*, Robert K. G. Temple's *The Sirius Mystery* (the mythology of the Dogon negroes) and Luis Navia's *Das Abenteuer Universum* (philosophical analyses).

Since 1968 the postman has brought me about 50,000 letters from readers. More than 43,000 newspaper cuttings are filed in my archives under persons and subjects. As the cutting service only supplies material from English- and German-speaking countries, the total number of articles would be more like 100,000. But even the predominance of thoroughly favourable stories has not quite removed the nasty taste left by a few malicious sensational articles, most of them full of von Däniken quotations which I had never uttered.

In 1972 the well-known lawyer Dr Gene M. Phillips, Chicago, founded the Ancient Astronaut Society*. He had

* Address in Europe: Ancient Astronaut Society, CH–4532 Feldbrunnen, Switzerland.

seen a cut version of my film *Chariots of the Gods* on American television. The idea of our planet being visited by the gods in prehistory so fascinated the jurist that he and some friends spontaneously decided to found a society to exchange ideas about theories and research. Gene Phillips wrote to me at the time, asking for my support.

In 1979 the AAS has 4,000 members in 42 countries. A good third of them are academics and nearly all the authors writing about 'my' speciality belong to it. Every year since 1974, the society has organised a world congress in a different country. The latest results of research are exchanged and communicated to the public in lectures and discussions (in which our critics also take part). World Congresses to date and for the future are as follows: 1974/Chicago, 1975/Zurich, 1976/Crikvenica (Yugoslavia), 1977/Rio de Janeiro, 1978/Chicago, 1979/Munich, 1980/New Zealand.

Our theory would not be so good as it is, if it had not aroused vehement criticism. Since 1968, 25 books *opposing* the Ancient Astronaut idea have been published. Nineteen of them claim to be 'scientific' either in the title or the introduction, although only 9 of the 19 were actually written by scientists. In spite of many express claims to be scientific, I have yet to see a genuinely 'scientific' book. This is sheer mislabelling, intended to exert a certain consumer pressure on the press. I admit that the critics' system is perfect. They all write more or less the same thing. For lack of real proof, the old familiar dud ammunition plops out of the sacred pages as 'counter-proof'.

I am entitled to call these examples of so-called counterproof duds, because in fact they prove nothing. This is the 'method'. If an archaeological book, whether by Heyerdahl, Ceram, Brion or Lhote, interprets finds from some site differently from me, I am refuted! If, contrary to prevailing dogma, I use modern knowledge to explain old texts differently, I am making a mistake. The hypotheses others put forward are sacrosanct, they are the ultimate wisdom and truth. If I hypothetically support a contrary, supplementary or ongoing opinion, I am in the wrong. It's as simple as that. How would things look today if our forefathers, too, had used this method to block every new progressive idea? Throughout our history certain authorities would dearly have loved to inscribe their point of view in marble as the ultimate in knowledge, so that any protest has always been a form of sacrilege and *lèse-majesté*. Those authorities would gladly have pilloried or better still burnt at the stake the opponents who were unwilling to swallow indigestible theories. The same holds good today. But if the rebels had accepted all the dogmatic views and doctrines as irrefutable truth, mankind would not have progressed in a single field of knowledge. Yet at all times progress can only be made if new views are put forward, indeed, must be put forward, because they are the mainsprings of development. To this compulsion to postulate new ideas we owe progress, development and the latest state of knowledge at any given time. Wernher von Braun (1912–1977), who ought to know, said:

> 'With hindsight, nothing looks so simple as a Utopia that has become reality!'

Although I am a burnt child, in 1977 I fell once again for promises. The producer Graham Massey visited me and charmed me into collaborating on an 'objective documentation' of my subject. As he came in the name of the BBC, which is normally so fair, I agreed to cooperate. My opponents, from Sagan to Heyerdahl, marched across the screen in serried ranks. I have nothing to say against that, but I should have been allowed to confront them. That would have been English fair play, but the negative statements appeared uncontradicted on the screen. It would also have been fairer if *supporters* of my theory had also been allowed to speak—man to man. Not a bit of it. My critics were the only ones to hold the floor.

This would not be worth mentioning, if Mr Massey's

'documentary' had not been screened in so many countries and if my dear critics everywhere had not swallowed the programme whole and then trotted it out against me in a 'scientific' manner: Däniken was 'unmasked' in the BBC documentary.

The well-known American astronomer Carl Sagan carried off the prize.

Since the end of 1977 a society that works contrapuntally, so to speak, has been active. It is called the Committee for the Scientific Investigation of Claims of the Paranormal. This committee is composed of 43 scientists, journalists and educators who want to kill off the 'new nonsense' in the USA (48). The leader of this organisation is Paul Kurtz, Professor of Philosophy at Buffalo State University. Need I say that Carl Sagan is one of the members? The committee is busy bombarding the press with ammunition against the ancient astronaut idea. It attacks television stations like NBC, one of the three major networks, which, in search of more objective information, allot space to people who think differently. The editors of popular magazines feel flattered by articles with academic dedications . . . and print them.

Marvellous. All this has happened since 1968. Caused by a single best-seller. A theory must indeed be incendiary and powerful to spark off such a 'battle' on stage and behind the scenes! I am very pleased. Surely it speaks highly for our society that a single inspired idea can set it rocking. Does it not show that beyond cars, refrigerators and other material comforts it still has a mind open to questions above the material level? That it is interested in the origin of man and that its 'dream' of the future is not exclusively occupied with the gross national product?

Although it is always enjoyable, I do not want to take the arguments of my friends on the other side one by one. I have done that in 'cross-examination'. But I must single out one point from the heated argument, because it is subjective and insidious. Especially in schools (as I know from the many letters I receive from schoolchildren) and publications aimed at the younger generation, it is subliminally or openly claimed that the gods=astronauts theory is harmful, indeed that it constitutes a danger to mankind. How is this done?

There are three basic claims:

1 There is no need for extraterrestrial visitors in prehistoric times in the accepted world picture. All the puzzling phenomena in the past can be explained more naturally and logically, and above all more simply, than by extraterrestrials visiting the earth and helping its inhabitants.

2 Supporters of the gods=astronauts theory label our early ancestors as stupid and limited, saying they were incapable of thinking independently or erecting monumental buildings without extraterrestrial help.

3 The theory is dangerous to mankind, because it induces man to believe in extraterrestrial gods, to hope for their help and so to sit back and wait for them to solve his problems.

These imputations demand an unequivocal refutation. They are indeed essential, vital points in the worldwide discussion. They act like drugs, crippling the brain and inhibiting thought.

How does the gods=astronauts theory look in reality? To take point 1:

I know no other theory which fits so perfectly and logically into our prehistoric past and so provides an explanation of the unsolved phenomena of those early days:

—The origin of life on earth.

—The origin of intelligence on earth.

—The difference between apes and intelligent man (the missing links).

—The identical protein structures in man and chimpanzee (the missing evolutionary driving force).

—The beginning of religions.

—The original core of global mythologies.
—Descriptions of God in the Old Testament accompanied by fire, quakes, din and smoke—as in many other ancient texts.
—The origin of giants and races.
—The list of the names of the fallen sons of heaven in the Book of the Prophet Enoch.
—The problem of God and the Devil, the primordial symbols of good and evil.
—The descriptions of divine punishment tribunals in prehistoric times.
—The worldwide Deluge.
—The legendary antediluvian kings and the patriarchs.
—Religious and mythological figures vanishing 'into heaven'. The origin of and/or motivation for hitherto unexplained buildings in prehistoric times (built out of respect for the 'gods', often erected with tools provided by the 'gods' or constructed with priestly knowledge from the 'divine' past).
—Shelters built as protection against the 'gods' (underground cities, inhabited cave labyrinths, dolmens).
—The effect of time dilation, constantly recurring in ancient texts (described in the Japanese Nihongi, and of the temporary disappearance of Abimelech in the Book of Baruch, etc.).
—The *first* mummifications (men hoped for physical rebirth when the gods returned.).
—Frequent mention of fear of the return of the gods (because man had transgressed against 'divine' prohibitions, he was afraid of punishment by the extraterrestrials.).
—The earliest sacrificial gifts to propitiate the gods (the extraterrestrials often accepted payment in kind for their 'evolutionary aid').
—The origin of certain foodstuffs described in mythologies, such as wheat and corn.
—The origin of ancient religious symbols and cults (the

cult of the sun, the cult of the stars, flying chariots in the heavens, wheels in the firmament, technical machines like the ark of the covenant and Solomon's flying cart). The origin of gigantic figures incised in the ground and laid out so that they could be seen by flying 'gods'). The origin of traditions (for example, Archangel Lucifer with the fiery sword fighting with Archangel Gabriel). The origin of countless religious inspired rock drawings all over the globe.

—The origin of religious and divine figurines in early antiquity (depictions of gods in helmets, figures wearing garments like spacesuits, gods with wings and technical accessories, etc.).

—The origin of cults which are still practised today in honour of the extraterrestrials (among the Kayapo Indians in Brazil or the Red Indian Hopi in Arizona, USA).

Obviously that is not a complete list. I merely want to remind you of a few vital points. If they were really objective, my critics would have to admit that these links fit exactly into the gaps which have hitherto existed in the early history of mankind.

The bald statement that 'extraterrestrials' are not necessary to illuminate the dark epoch of our past and that the theory of their former presence on our planet could not explain anything would be untenable, given a little impartiality by the other side.

So where are the 'simpler' answers to the unsolved puzzles of the past? Is the gods=astronauts theory ultimately refutable, because it supplies answers that *really are simpler*? Is it 'simpler' to assume that the evolution of man to the stage of *homo sapiens* was due to a millionfold chance in genetic evolution than to admit that extra-terrestrials created intelligent beings 'in their own image', as traditions tell us? Is it not sheer mischief, rather than a simple answer, to claim that the origin of early mythologies and religions (with the technical data often given in their texts) is more plausibly explained by psychological claptrap? If you accept (just as an experiment, please) the former presence of extraterrestrials, there is no need to strap our ancestors down on the psychiatrist's couch to extract the vaguest of explanations from them. Admittedly it is simpler to deny the existence of giants in prehistory than to come to grips with the phenomenon. The traces of giants in ancient texts simply cannot be overlooked, any more than the prints they left in the ground when they were on earth—prints that have been photographed. One cannot call this method a simple answer to a difficult question.

Occupational blindness occurs everywhere, and not only in scientific circles, once disturbing new knowledge is deliberately disregarded. The 'experts' prefer to snatch at the most absurd explanations rather than pay the least attention to anything new, so that they can go on contemplating their own navels in the shrine of inherited knowledge. 2,500 years ago, the gods said to our ancestor, the prophet Ezekiel: 'Thou dwellest in the midst of a rebellious house, which have eyes to see, and see not.' Today they would certainly add: 'They have reason, and use it not!'

Now to point 2. I never wrote that our ancestors were stupid and incapable of erecting prehistoric buildings. I never said that extraterrestrials built megalithic temples or pyramids, or drew the pictures on the plains of Nazca. Those are malicious assertions by my prejudiced opponents.

However, I do support the view that the *reason and motive* for the construction of some mysterious buildings can be traced back to extraterrestrial beings or that our ancestors used building techniques in which they were instructed by the 'gods'. There are good grounds for my assumption. How else can one explain the identical masterpieces visible all over the world? According to contemporary doctrine, the various early cultures developed quite independently of one another—on Easter Island and in Brittany, among the pre-Incas or the inhabitants of Great Britain (Stonehenge). Or anywhere else you care to mention. Yet that cannot have been the case.

When I see stonework on Easter Island that exactly resembles the stonework above Sacsayhuaman (Peru) and when I find the same 'manufacture' in Malta, Catal Hüyük (Turkey) and at Baalbek (Lebanon), the question insistently arises: where was the international school of stonemasonry which sent mastermasons all over the world to use the same building technique? After all, there was no organised communication by plane or ship, no magazine called *Megalithic Building Today*.

My simple assumption.

If our ancestors piled up gigantic monoliths to form temples and pyramids all over the world, without knowing about one another, there *must* have been a *common* motive for this prodigious slave labour.

When the prophet Enoch dishes up several pages of astronomical titbits (the meaning of which he cannot possibly have understood in his day) and also claims that they were all dictated to him by the watchers of the heavens, it must be admissible to ask who these watchers of the heavens really were.

Even if such traditions about what our past was like leave questions to answer, I still do not call our ancestors stupid. On the contrary, I think they were highly intelligent. They adapted themselves to progress, which is more than I can say about some critics of my theory.

When scholars speculate about the IQ's of our ancestors, the claim that experiments have shown that monkeys, especially chimpanzees, possess creative intelligence, keeps on cropping up. In a series of experiments monkeys have been trained to press special buttons to get to their food and water, to turn on the light or to use a switch to open the communicating door to the next cage.

On October 11, 1978, the *Frankfurter Allgemeine Zeitung* reported:

'A city for 10,000 monkeys is now being built near Adler on the Black Sea. The monkeys belong to the Scientific Research Institute for Experimental Pathology and Therapy. As the Moscow weekly *Nedelya* reported, architect Vadim Adamovich's plan envisages a laboratory and the huts necessary to house the animals, on a site of more than 84 hectares. Each hut will have running water, a bamboo bed and be in daylight. The walls will be painted in epoxyd colours. Each hut will have its own open-air enclosure. According to *Nedelya*, there has never been such a large enclosure for monkeys. A kind of fence of spiked iron arches will separate the monkey city from the rest of the world.' (49)

If we receive a report on this project in a few years' time, we shall learn to our astonishment that our charming ancestors have acquired a certain limited intelligence. That they carry on like human beings in their bamboo marriage beds, that they no longer eat bananas with their hands, but cut them up and put them in their mouths with a knife and fork, that they use the WC most hygienically and communicate from cage to cage by telephone.

The *Schweitzer Illustrierte* describes the laborious training needed to make monkeys house-trained. (50)

Coco, a female gorilla, was born in San Francisco Zoo on July 4, 1971. A young lady called Penny Patterson took Coco over and after *seven years* of round the clock communal life taught her 350 words with which she can now make her wishes known. Teacher Penny first learnt the deaf and dumb language so that she could communicate with Coco. Now, after seven years, Coco uses the signs. Recently Coco was given a partner, a male gorilla called Michael. It remains to be seen if the children will inherit their mother's intelligence or whether Miss Patterson will have to take them to kindergarten again. After all, she only trained one monkey; she did not change the whole tribe!

The intelligence of chimpanzees cannot be advanced very far. There is a Monkey School (51) with a Chimpanzee Rehabilitation Programme on a Senegalese game reserve. Since 1968, Stella Brewer has been taking in chimpanzees which have lost their parents or fallen under human influence in zoos or circuses and teaching them what they need to know to live and survive in the freedom of the jungle. Back to nature—after human instruction!

So monkeys acquire a little intelligence through human effort and teaching. Man teaches them to press certain buttons, to say words, to understand signs and how to live in their own environment.

None of this occurred to the monkeys of their own accord. *We human beings* were their teachers. If these trained chimpanzees and gorillas should reach a certain independence in a few generations, if they could speak even halfway 'rationally' and so develop a sort of civilisation, *we human beings* would have played the role of 'gods' in their existence. *We* handed on knowledge, *we* gave them the foundations on which they developed. From the point of view of 'thinking' apes *we* had the intelligence and power. So in my view the experiments prove exactly the opposite of what their initiators had in mind. It was *not* proved that apes were independent from time immemorial, but rather that they could become so with outside help!

I leave it to my readers to draw their own conclusions from this excursion into the beginnings of human intelligence. Who were *our* teachers?

Finally, point 3. Is the gods=astronauts theory dangerous? Can it 'seduce' people into waiting idly for life to go by, into hoping that extraterrestrials will solve their problems? That is the most narrow-minded of all insinuations! Anyone who supports this lie should attack the representatives of established religions which promise 'help from above'. What about this consoling thought: 'The Lord will provide?' What do children learn at the Sunday Schools of the major religious groups and sects? 'Knock, and it shall be opened unto you.' 'Man proposes and God disposes.' 'Ask and it shall be given unto you.' 'Blessed are the poor in spirit: for theirs is the kingdom of heaven.'

The danger of denying fate, of underestimating one's own powers or leaving decisions to some ill-defined being, has no part in the gods=astronauts theory, quite apart from the fact that it is not, nor does it set out to be, a doctrine of salvation. And yet I claim that the extraterrestrials will return!

Christians await the return of the Lord. In the Gospel it says that he will come with great power and glory to sit in the clouds and judge.

This hope of return existed over 2000 years ago when Jesus was among the Jews, who had long been awaiting their Messiah. They did not recognise Jesus as their saviour.

In the Old Testament there are figures like the antediluvian prophet Enoch who disappeared for ever with the watchers of the heavens or his colleague Elijah of whom it is said that he vanished into the clouds in a fiery chariot. According to traditional teaching, Enoch and Elija should return to die here.

Before he took off into the cosmos, Bep-Kororoti promised to return one day. He was the 'warrior from space' of the Kayapo Indians on the Rio Fresco in Brazil and they worshipped him as a god. The Kachina, gods of the Red Indian Hopi in Arizona, also promised to return when they said goodbye.

When the white conquerers (1542/25) visited the Inca empire they were greeted joyfully on landing, because tradition promised that the gods would come back. In their religious innocence, the Incas mistook the Spanish hordes under the gold-crazy Francisco Pizarro for their homecoming gods. The Aztecs in Central America made the same mistake. When Hernando Cortez besieged Tenochtitlan, then the biggest city in America, in 1519, its capture was made much easier because the Aztecs thought the conquistador was their long-awaited god. When the navigator James Cook discovered the Hawaiian islands in 1778, the natives took him for their golden-haired god Lono who was returning to his country.

What kind of gods were these who made binding promises of the sort so frequently recorded in ancient traditions? They cannot have been nebulous spirits or imaginary phantoms. They were physical beings who came from heaven and lived among our ancestors—beings so superior that the power of gods was ascribed to them. When they set off heavenwards again, their promise to return was taken as a matter of course.

Was the extraterrestrials' promise to return a frivolous one? No. They knew the physical law of time dilation, according to which astronauts in spaceships travelling close to the speed of light are subject to a different time from that on earth. They knew perfectly well that only a few years would pass for them in their spaceships, whereas millennia would speed by on earth. The extraterrestrials were able to make a promise to return, because they will fulfil it!

What do the extraterrestrials expect when they return?

A planet on which peoples fight each other to uphold their own stupid dogmas? Easy-going idle men who have neglected, forgotten or misused the 'divine' heritage of intelligence? Do the gods expect a planet with an advanced technology, atomic powerstations and spaceships, or a Stone-Age culture with men sharpening arrows by the light of oil lamps in dismal caves. Do they expect a society whose members covet the property of others or a society with a firm moral and ethical foundation which keeps the commandments imposed on it.

Which commandments?

The path and goal of human endeavour are most clearly laid down in the Old Testament:

'And God blessed them, and God said unto them, Be

fruitful and multiply, and replenish this earth, and subdue it: and have dominion over the fish of the sea, and over the fowl of the air, and over every living thing that moveth upon the earth.'

Genesis 1,28

The task is plain. We are to use our human intelligence to dominate the animal kingdom and 'subdue' the earth with all its riches, oxygen, water, minerals, oil, etc, which are at our disposal.

It makes no difference whether we look on the Old Testament God who gave the instruction as some inconceivable omnipotent spiritual being or an extraterrestrial figure: the God or 'Gods' were far superior to men. He or they knew what the imperative 'multiply' would lead to, namely overpopulation and so to wars for new territories, to a shortage of food and clothing, in short to a state of necessity which could only be over-come by intelligence. That is why they endowed men with the intelligence which enabled them to solve their problems. That is how we should understand the promise contained in this sentence:

'. . . The people is one . . . and this they begin to do: and now nothing will be restrained from them, which they have imagined to do.' Genesis 11, 6

And as we possess the intelligence to overcome our problems, we have no reason to sit twiddling our thumbs and waiting for extraterrestrial help!

In the Bible God is spoken of in the singular. Is it an inadmissible trick for me to speak of 'Gods'?

In the original Hebrew text the plural concept 'Elohim' stands for 'God'. The verb in front of the plural concept is in the singular. For example: 'And the Gods created men in their own image.'* As the verb is in the singular, the translators changed the plural concept 'Elohim' into the singular, too, making it 'God'. But expert theologians

*The German *schuf* = created is in the singular, but this, of course, is lost in English, Translator's note.

assured me that it would be equally admissible to assimilate the singular *schuf* (created) to the plural of 'Elohim'. Then the translation would read: 'The Gods *schufen* (German plural of created) men in their own image.'

How could the 'Gods' order men to multiply and subdue the earth when they must have known that the consequences of following out their orders would be disastrous?

The 'Gods' left early mankind with clear commandments. If they were kept, there would be an intact smoothly running civilisation with an assured future and a high level of culture. The ten commandments the gods imposed on our ancestors can be found in Exodus, chapter 20, verses 2–17, and in Deuteronomy, chapter 5, verses 6–21. In translations of the Bible each commandment is preceded by the imperative: 'Thou shalt not.' Actually the translation 'Thou wilt not' would be equally admissible, for the Hebraic concepts cover both possibilities.

If we consider certain commandments as being inspired by a prophet of the past, they acquire a new aspect throwing light on the future return of the gods.

The first commandment says:

> 'Thou shalt not make unto thee any graven image, or any likeness of anything that is in heaven above, or that is in the earth underneath, or that is in the water under the earth.'

The extraterrestrials knew perfectly well that they were *not* omnipotent immortal gods. They, too, probably worshipped the inconceivable something that, for want of a better word, is called 'god' in all religions. They also knew that our innocent ancestors took them for 'gods', so they tried to make a clear division between themselves and the inconceivable god. To prevent later generations worshipping idols of stone, wood and plastic, they imposed a strict ban on making images of 'God'. And what happened? Immediately after the extraterrestrials took off into the depths of space, men disregarded the commandment. Every religion in every civilisation made images of the gods and christened them with a variety of names. The only religion which kept the commandment that I know of is Islam, which will not tolerate any kind of divine image.

The fourth commandment says:

'Honour thy father and thy mother: that thy days may be long . . .'

That is surely the only one of the Ten Commandments which is kept by all civilised peoples.

I find the promise of longevity coupled with the commandment interesting because it has a present-day significance. Why should someone who honours his parents live longer? Isn't this a hint at something which was first confirmed by modern research? Namely that the warmth of the parental home gives the psyche an almost lifelong sense of peace and security, just as—vice versa—a psyche damaged in youth makes a man unhappy. Men with psychological troubles, according to modern research, are more susceptible to cancer. Therefore keeping the commandment promises a longer life expectancy.

Commandments five to eight are simple and clear, and if strictly followed would bring heaven on earth:

Thou shalt not kill!

Thou shalt not commit adultery!

Thou shalt not steal!

Thou shalt not bear false witness against thy neighbour!

That is really the simple formula for peace and happiness. How wonderful it would have been if men had always followed the intelligent laws of the 'gods'! A world in which there was no killing on any pretext. No wars, no genocide. The television news which brings us nothing else every day could say something cheerful for a change. Yes, peace *really* was programmed thousands of years ago.

'Thou shalt not commit adultery . . .' I feel that this commandment did not count for much even at the time it was issued. Intoxicated with boundless freedom, people nowadays don't want to know about it. How much strife and misery would have been avoided, how many tears would not have been shed, if this primordial law brought from another star had been obeyed!

We should have to shout halleluja if the commandment 'Thou shalt not steal' had been kept. No locks on the doors, policemen pensioned off, bags and pockets without fastenings . . . because there is no stealing! But reality has turned a good commandment into an unachievable Utopia.

'Thou shalt not bear false witness . . .' How many millions of men—since men have been able to think—have been sentenced on the basis of false witness? How many millions of times have men told lies about their neighbours. An intelligent law which is disregarded daily, but that does not make it a bad law.

The ninth commandment seems to have been formulated from a great wealth of experience:

> 'Thou shalt not covet thy neighbour's house, thou shalt not covet thy neighbour's wife, not his manservant, nor his maidservant, nor his ox, nor his ass, nor anything that is thy neighbour's.'

How farsighted the 'gods' were! How well they knew the beings they had endowed with intelligence! Envy, they knew, was the destroyer of all communal life.

The trail of poisonous envy runs through the history of mankind, broadening as it goes. Is it not persistently propagated—both secretly and openly—in the universities by many interested parties? What other people have acquired by hard work must be redistributed! When all attendant phenomena are removed, is not envy the real reason for the major disputes between peoples and individuals? With their knowledge of evolution, the gods knew what laws to give and why.

Far from thinking in an arch-conservative way, worlds apart from those who infinitely prefer the past to the present (and the future), I assert that our planet would be close to paradise if men had stuck to the simple commandments of the 'gods'. Those ancient commandments comprise every prerequisite for a flourishing communal life. Without ifs and buts. There is no set of laws in the world, however cleverly worked out, that provides anything like such a convincing code in so few succinct commandments and prohibitions.

I only hope that the gods will not make an inventory tomorrow and realise what has become of their grandiose plan.

So what do the extraterrestrials expect when they return? How will they react to the state of our society, with its dubious achievements?

The coordinates of our solar system are stored in the computer on board their spaceship and the goal, our planet earth, is also programmed on it. It makes no difference whether the original crew, only a few years older, or a new generation fulfils the promise to return. It is rather like our modern colonisers who instructed and helped 'the natives'. They allotted praise and thanks if their orders were adequately carried out, or they took Draconian measures if they were not.

If anyone retorts that extraterrestrials would never behave like colonials, indeed, that the very idea is a product of reactionary thinking. I must convince him of his error. The 'gods' created human intelligence 'in their own image'. That is why *our* thought processes are very similar to those of our divine ancestors. Tradition tells impressive stories which show that the 'gods' did not treat the inhabitants of earth with kid gloves in prehistoric times. They unceremoniously destroyed whole cities with fire and brimstone that rained down from heaven. Furious with their brood, they had no hesitation in drowning most of mankind, as described in the Sumerian Epic of Gilgamesh and the biblical account of Noah and the Flood. We must ask ourselves whether extra-terrestrials could not behave equally rigorously in *our* day.

Time can be manipulated by energy. Anyone who has unlimited supplies of energy at his disposal can achieve anything, for *time* is on his side. If the crew of an extraterrestrial spaceship destroyed present-day human civilisation by a shower of bacteria, they could wait until human intelligence regained a certain degree of civilisation. The extraterrestrials possess enormous amounts of energy; they can board their spaceship and fly to another solar system. While they grow only a few years older, 10,000 years may pass on our planet, depending on the ship's velocity. When they return, a new civilisation has developed. That is why the extraterrestrials can permit themselves to remove unsatisfactory products of their 'colonial activity' from time to time. For time is on their side.

What can we do to avert the wrath of the returning 'gods'? Have we any chance of catching up with them, so that we need no longer fear their superior technology?

The original order said that we should 'subdue' the earth. Man was told to 'Grow and multiply'. (German version of Bible.) 'Grow' and 'multiply' do not have the same meaning, i.e. they refer to two separate 'godgiven' orders. We understand growing as getting bigger, multiplying as reproducing on a large scale.

If extraterrestrials implanted their intelligence in us, 'growing' would refer to the growth of intelligence, and all intelligence finds the driving force to grow bigger in curiosity—especially scientific curiosity. Inspired by the intelligence of the superior extraterrestrials, scientific curiosity is confirmed in the discovery of the resources of energy on our planet.

I know the hoarse-voiced male choir which compassionately warns us against plundering our planet. It grieves me deeply how little they think of human intelligence. For it will always be ingenious enough to replace dwindling raw materials by others. The sum is simple. Rare raw materials are dear, and the rarer they are, the dearer they become, until one day they are priced out of the market. At this moment at the latest, intelligent man remembers that he can achieve the same effect with another material. Man will always find *the* way. All the engines in the world which run on petroleum and its products could be switched to hydrogen today. Born of the necessity which the proverb says is the mother of invention, recycling processes are popular. Something different or new can be made out of nearly every kind of refuse.

The ship's surgeon Dr Robert Mayer (1814–1878) discovered the law of the conservation of energy. This states that the total energy in the universe is constant and that all forms of energy are convertible into one another. Wernher von Braun wrote:

> 'Science has established that nothing can vanish without a trace. Nature knows no destruction, only change.'

Whether it came from the one God or my extraterrestrial gods, the charge laid on men always had the same meaning. They had to subdue the earth and grow, to emulate God or the 'gods'. So it would be like committing hari-kiri to turn our ancient mission upside down: to calumniate the advance of technology, to impede the exploitation of our resources, to leave atomic power untapped. The divine mission entrusted to us was quite different.

Mankind should prepare itself morally, ethically *and* technically for the return of the 'gods'. The Ten Commandments, the expression of perfect wisdom, should be enthroned again. Our intelligent curiosity should be given the value the 'gods' ascribed to it. With this ultra modern programme hunger could be banished from the world, wars would be a miserable spectre from the past and significant work would not be a dream of Utopia. Extraterrestrials will only accept us as their partners if we at least resemble their image. Given this creed, can anyone honestly claim that there is anything in the far-reaching gods=astronauts theory that wants or encourages people to stand idly by, awaiting the help of the 'gods'? If the theory were grasped in its constructive, positive aspect, mankind could confidently move towards a peaceful future, blessed with progress. It would no longer have to fear the return of the 'gods'. Nevertheless the state in which it has to show our planet does not inspire confidence. It is up to us to change it.

To quote the Olympian J. W. Goethe:

'We are used to men deriding what they do not understand.'

Communiqué

In its last number of the year, No 52/1978, *Der Spiegel* included a fourteen-page report entitled 'Astronomy: a new scenario of the cosmos'.

Careful readers of my books met many 'old friends' in this basic compact article.

To end this book I want to quote in full an article from *Der Spiegel* (No 1/1979), because of its really remarkable contents. It makes me hope that the claim I have repeatedly made in my books and lectures that all the relevant branches of science should take on the search for extraterrestrial intelligences has at last found an echo:

> *Cosmic search*
>
> '"I cannot imagine a more frightening nightmare than communication with a so-called superior . . . civilisation in space." This quotation from the Harvard biologist and Nobel Prize winner George Wald comes from the first number of a new periodical which is devoted to the search for alien intelligences, *Cosmic Search* will appear bimonthly as from January at an annual subscription of 16 dollars outside the USA and should be taken seriously. The editorial offices are at the Radio Observatory of Ohio State University and contributors include such serious scholars as the British astronomer Martin Rees of Cambridge University, Nicolai Kardachev of the Space Research Institute of the Soviet Academy of Sciences and John Billingham, Director of the American Search for Extraterrestrial Intelligence programme (Seti). Nothing then to send UFO freaks on a trip, but good sound stuff for earthlings who miss a bit of fantasy in science.'

Bibliography

(1) *Pierers Konversations-Lexikon*, Vol. 3, Berlin, 1889.

(2) GRESSMANN, HUGO, *Die Lade Jahves und das Allerheiligste des Salomonischen Tempels*, Leipzig, 1920.

(3) SCHMITT, REINER, *Zelt und Lade als Thema alttestamentlicher Wissenschaft*, Gütersloh, 1972.

(4) DIBELIUS, MARTIN, *Die Lade Jahves—Eine religionsgeschichtliche Untersuchung*, Göttingen, 1906.

(5) VATKE, R., *Die biblische Theologie—wissenschaftlich dargestellt*, Berlin, 1835.

(6) TORCZYNER, HARRY, *Die Bundeslade und die Anfänge der Religion Israels*, 1930.

(7) *Neues Theologisches Journal*, Nuremberg, 1898.

(8) SASSOON, G. AND DALE, R., *The Manna Machine*, London, 1978.

(9) *Realencyklopädie für protestantische Theologie*, Vol. 8, Leipzig, 1900.

(10) LANDE-NASH, IRENE, *3000 Jahre Jerusalem*, Tübingen, 1964.

(11) KAUTZSCH, E., *Die Apokryphen und Pseudepigraphen des Alten Testaments (Reste der Worte Baruchs),* Tübingen, 1900.

(12) *Theologische Studien und Kritiken*, No 1, Gotha, 1877.

(13) *The Holy Bible*, Revised Version, Old Testament Apocrypha.

(14) *The Misnah*, translated from the Hebrew by Herbert Danby, Oxford University, no date.

(15) BLUMRICH, JOSEPH, *The Spaceships of Ezekiel*, Bantam Books, 1974.

(16) ENCYCLOPAEDIA JUDAICA, 'Das Judentum in Geschichte und Gegenwart, Berlin, undated.

(17) *Bulletin of the John Rylands Library*, Vol. 25/1, Manchester, 1962.

(18) *Abhandlungen der Philosophisch-Philologischen Klasse der Königlich Bayerischen Akademie der Wissenschaften*, Vol. 23, Section 1, Kebra Negast, Die Herrlichkeit der Könige. Also

used *The Queen of Sheba and her only son Menyelek, (Kebra Nagast)*, translated by Sir E. A. Wallis Budge, London, 1932.
(19) SCHMID, JAKOB, 'Vom Gebirgsland Semien zum Roten Meer', *Neue Zürcher Zeitung*, 24.10.1970.
(20) SUDHOFF, K., Paracelsus, Collected Works, Vols. 4–6, 1923.
(21) RORVIK, D. M., *In his Own Image*, London, 1979.
(22) PAUWELS, LOUIS, *Manifest eines Optimisten*, Berne, 1972.
(23) 'Durch Gen-Rutsch zum nackten Affen', *Der Spiegel*, 18/1975, Hamburg.
(24) OPPENHEIMER, J. F., *Lexikon des Judentums*, Gütersloh, 1967.
(25) SPEICHER, GUNTER, 'Mensch und Tier aus der Retorte', *Welt am Sonntag*, 23.7.1978., Hamburg.
(26) SCHULTZE, H., 'Die ersten geclonte Mäuse leben schon', *Frankfurter Rundschau*, 26.8.78.
(27) 'Was die Welt bewegte, schwappt im Reagenzglas', *Der Spiegel*, 36/1978, Hamburg.
(28) *Lexikon der Prä-Astronautik*, Düsseldorf, 1979.
(29) TICHY, HERBERT, *Rau-Tau, Bei Gottern und Nomaden der Sulu-See*, Vienna, 1973.
(30) *Lexikon der Archäologie*, Reinbeck, 1975.
(31) HERM, STEFAN, *Die Phönizier*, Düsseldorf, 1973.
(32) *Ancient Skies*, 5/1978.
(33) WIESINGER, J./HASSE, P., *Handbuch fur Blitzschutz und Erdung*, Munich, 1977.
(34) TOBISCH, OSWALD O., *Kult Symbol Schrift*, Baden-Baden, 1963.
(35) FEINBERG, GERALD, 'Possibility of faster-than-light particles', *Physical Review*, 1967.
(36) KIRCH, DIETMAR, 'Tachyonen—Teilchen schneller als das Licht', in *Wissenschaft und Technik*, Frankfurt, 23/1977.
(37) MANIAS, DR THEOPHANIS M., *The invisible harmony of the ancient Greek world and the apocryphal geometry of the Greeks—The geometric geodetic triangulation of the ancient Hellenic space*, Edition of National Institution, Athens, 1969.
(38) ROGOWSKI, PROF. DR FRITZ, 'Tennen und Steinkreise in Griechenland, Mitteilungen der Technischen Universität Carolo-Wilhelmina zu Braunschweig.' Published by Prof. Dr Edgar R. Rosen in cooperation with the Braunschweig Hochschulbund, VIII/2/1973.

(39) GRETHER, EDWARD, Theorieheft Planematerie, Part 2, no date.

(40) BODNARUK, NIKOLAI, 'Das geheimnisvolle Netz auf dem Globus', from *Komsomolskaya Pravda, Sputnik*, 9/1974.

(41) BRION, MARCEL, *Die Frühen Kulturen der Welt*, Cologne, 1964.

(42) TEMPLE, ROBERT K. G., *The Sirius Mystery*, London, 1976.

(43) WOOLLEY, SIR CHARLES L., *Ur of the Chaldees*, London, 1929.

(44) PARROT, ANDRÉ, *Sumer*, London, 1960.

(45) CERAM, C. W., *Gods, Graves and Scholars*, London, 1971.

(46) SCHMIDTKE, FRIEDERICH, *Der Aufbau der Babylonischen Chronologie,* Münster, 1952.

(47) BONDI, HERMANN, *Die Wiege stand im Orient*, Munich, 1971.

(48) 'Attacking the new nonsense', *Time Magazine*, 12.12.1977.

(49) 'Eine Stadt für 10,000 Affen', *Frankfurter Allgemeine Zeitung*, 11.10.1978.

(50) 'Und was sagt der Affe dazu?', Schweizer Illustrierte, 45/1978.

(51) BREWER, STELLA, *The Forest Dwellers*, London, 1978.

GENERAL

ARRHENIUS, S., *Das werden der Welten*, Leipzig, 1907.

BENZINGER, J., *Hebräische Archäologie*, Leipzig, 1894.

BIBBY, GEOFFREY, *Looking for Dilmun*, London, 1969.

BLUCHEL, K., *Projekt Ubermensch*, Berne, 1971.

BONDI, H., *Die Wiege stand im Orient*, Munich, 1971.

BRACEWELL, R. N., *Life in the Galaxy*, New York, undated.

BRION, MARCEL, *Die frühen Kulturen der Welt*, Cologne, 1964.

CALDER, NIGEL, *Mind of Man,* London, 1970.

DILLMANN, A., *Handbuch der alttestamentlichen Theologie*, Leipzig, 1895.

GIOT, P. R., *Menhirs et Dolmens*, Chateaulin, undated.

KRAMER, S. N., *The Sumerians*, Chicago, 1962.

LAPP, R. E., *Kill and Overkill, The Strategy of Annihilation*, New York, 1961.

Index

(Illustrations are shown in italics)